Andrew Whyte Barclay

Gout and rheumatism in relation to disease of the heart

Andrew Whyte Barclay

Gout and rheumatism in relation to disease of the heart

ISBN/EAN: 9783744736930

Printed in Europe, USA, Canada, Australia, Japan

Cover: Foto ©berggeist007 / pixelio.de

More available books at **www.hansebooks.com**

GOUT AND RHEUMATISM

IN RELATION TO

DISEASE OF THE HEART.

BY THE SAME AUTHOR:

A MANUAL OF MEDICAL DIAGNOSIS; being an
Analysis of the Signs and Symptoms of Disease. *Second Edition.*
Fcap. 8vo. 8s. 6d.

MEDICAL ERRORS: Fallacies connected with the
Application of the Inductive Method of Reasoning to the Science of
Medicine. Post 8vo. 5s.

GOUT AND RHEUMATISM

IN RELATION TO

DISEASE OF THE HEART.

BY A. W. BARCLAY, M.D.,

CANTAB. & EDIN.;

FELLOW ROY. COLL. PHYS.; PHYSICIAN TO ST. GEORGE'S HOSPITAL, ETC.

LONDON:

JOHN CHURCHILL & SONS, NEW BURLINGTON STREET.

MDCCCLXVI.

LONDON: PRINTED BY W. CLOWES AND SONS, STAMFORD STREET,
AND CHARING CROSS.

PREFACE.

In submitting to the judgment of my medical brethren the lessons of experience on the questions which form the subject of the following pages, I may remind them that it is one which has long occupied my thoughts, not, indeed, to the exclusion of other departments of medical science, but with an interest which early investigations alone inspire when new and untried paths of knowledge first open to the mind of the student. At a period, when I held for several years in succession the appointment of Medical Registrar at St. George's Hospital, it seemed possible by a careful and exact inquiry into the clinical history of the very large number of cases passing constantly

under review, to learn something more than
could be found in books, of the pathological
relations of a class of diseases so important
and so little amenable to treatment. The
results of my observations were submitted to
the Royal Medical and Chirurgical Society,
and were published in the xxxist and xxxvth
volumes of their Transactions; and they afterwards formed the ground-work of the rules for
diagnosis which have been laid down in my
'Manual of Medical Diagnosis' for the use
of students.

In again returning after so long an interval
to the same subject, my object is rather to
complete what has been already begun than
to add any new ideas : to substitute the
teaching of experience for the more crude
theories of early inquiry. I have been in-
duced to arrange my thoughts in their present
shape rather than add another to the already
existing monographs on disease of the heart,
because the association of organic lesion with
rheumatic fever is a pathological fact of

universal application. This association was
the first subject of my own studies in cardiac pathology, and it is in the course of
rheumatic fever that the early symptoms of
disease may be traced, and the significance
of the simplest signs most easily learnt. The
plan of this work has led to my saying
something of the rheumatic disorder itself
beyond its mere relation to disease of the
heart, and also of the contrast afforded by
gout—at once so similar to, and so unlike
rheumatism.

On the subject of gout, I know that my
views are not entirely in harmony with those
of some who have contributed very much
to our present knowledge; but there are
others, and among them stands prominent the
respected name of Dr. Gairdner, with whose
opinions my own are much more in accord-
ance. Its treatmen does not occupy a prominent place in this volume, as the whole
management of such cases may be so much
better learnt from works more especially de-

voted to its consideration. I may, however, add that since these pages have been in type, I have been tempted to try the new French specific, to which the gouty are at present so constantly having recourse. Numerous reports of its employment had reached me, and if in one or two instances dangerous symptoms had arisen, it seemed in the majority to have been taken with impunity. The result of my own trials has not inspired me with any confidence in the virtues of the remedy, as it has failed in cases which have readily yielded to other means. When purgative action can allay the acute symptoms, it may no doubt do good, and this property will probably render its constant use more harmless than that of many other so-called specifics. It is not improbably a combination of some remedy of this class with a purgative.

The study of gout and rheumatism has not yet reached such a point that we may venture to speak of our knowledge regarding them as philosophical in its higher sense.

That some of the facts have formed the basis of true inductions is unquestionable; but much remains which as yet has no element of certainty. In such circumstances, I have thought it not unreasonable to venture on theoretical views which are certainly open to criticism. Let it only be remembered that they are put forward as hypotheses which seem to serve as an explanation of the facts observed, and that nothing more is claimed for them: the judgment of others must form the only true estimate of their value.

Bruton Street,
May, 1866.

CONTENTS.

CHAPTER I.

THE ORIGIN OF GOUT.

CHAPTER II.

ACUTE RHEUMATISM IN RELATION TO DISEASE OF THE HEART.

CHAPTER III.

RHEUMATIC INFLAMMATION OF THE HEART—ITS SYMPTOMS AND PROGRESS.

CHAPTER IV.

TREATMENT OF RHEUMATIC FEVER AND INFLAMMATION OF THE HEART.

CHAPTER V.

THE RELATION OF GOUT TO CARDIAC DISEASE.

CHAPTER VI.

DISEASE OF THE HEART IN GOUTY SUBJECTS.

CHAPTER VII.

THE GOUTY PAROXYSM AS IT AFFECTS THE HEART.

CHAPTER VIII.

GENERAL VIEW OF DISEASE OF THE HEART.

CHAPTER IX.

THE TREATMENT OF DISEASE OF THE HEART.

CHAPTER X.

TREATMENT OF GOUTY AFFECTIONS OF THE HEART.

RULES FOR THE DIAGNOSIS

OF

CARDIAC MURMURS.

1.—THE rhythm of the heart's action must first be ascertained by comparison of—

(*a*). The apex-beat felt with the finger.

(*b*). The sound heard over the centre of the heart.

(*c*). The sound heard above the base of the heart.

2. — *Bruits* are either (*a*) rhythmical (corresponding in rhythm to that of the heart's sounds), or (*b*) non-rhythmical (having a rhythm of their own, differing from that of the heart).

3.—A non-rhythmical bruit is always external to the heart —pericardial friction, pleuritic friction, &c.

4.—Rhythmical murmurs are either systolic or diastolic; and are heard at the base of the heart, over its centre, or at the apex.

5.—A systolic murmur at the base is either functional or indicates disease of the aortic valves; very rarely of the pulmonic valves.

(*a*). The functional murmur is diffuse, and is best recognised by the existence of a similar murmur in the carotid, or a " bruit de diable" in the veins of the neck.

(*b*). An aortic murmur is heard loudest close to the edge of the sternum, and is occasionally propagated along the aorta.

(*c*). A pulmonic murmur is heard locally loudest at the base of the heart about two inches from the sternum.

6.—A systolic murmur heard over the centre of the heart is either functional or reflected.

(*a*). Functional murmur may be heard loudest in this situation. It is diffuse, not local, in character.

(*b*). A reflected murmur may be traced more distinctly at some other part.

7.—A systolic murmur at the apex is generally mitral, occasionally functional, rarely from disease of tricuspid valves.

(*a*). The murmur of mitral regurgitation is remarkably local, sharp, and distinct.

(*b*). The functional murmur at the apex is diffuse and indistinct.

(*c*). The tricuspid murmur, when heard, is at some distance from the apex and near to the sternum.

8.—A diastolic murmur at the base indicates regurgitation through the aortic valves.

9.—A diastolic murmur over the centre of the heart is the same sound carried a little further down. In both there is very often also a systolic aortic murmur.

10. A diastolic murmur at the apex is usually also dependent on disease of the aortic valves; very rarely on disease of the mitral.

(*a*). When a systolic murmur is heard at the base, the murmur is unquestionably aortic; but when there is no roughness of the valve, it will be heard with the diastole only.

(*b*). A diastolic mitral murmur is only to be heard when there is also a mitral systolic murmur. It is otherwise necessarily dependent on aortic regurgitation. The mitral diastolic is exceedingly rare.

11.—Friction-sound really differs in quality as well as rhythm from endocardial murmurs; but this indication is not to be absolutely relied on.

12.—When there is any degree of obscurity it is well to cause the patient to hold his breath once or twice during the examination. This circumstance will not at all alter endocardial bruits, but sometimes affects pericardial friction, and will put an entire stop to pleuritic friction or any other sound produced in the lungs.

CHAPTER I.

THE ORIGIN OF GOUT.

Theories of Gout — Uric Acid in its relation to Gout — The chemical theory not tenable — Hereditary transmission favoured by Excess — Probable changes in the Blood — The relation of Gout to Kidney Disease — General views of Secretion — Secretion of Uric Acid — Various effects of excessive stimulation — Theory of Blood-change — Its permanent character — Probably due to Blood-cell growth — Influence of treatment.

THE great advances which have been made of late years in our knowledge of pathology and skill in diagnosis have given us a degree of certainty in estimating the existence and extent of internal disease, which can hardly have been guessed at by those who lived just before the era which has been so remarkable as one of experimental investigation and inquiry. The unpractised ear may indeed be still deceived by the sounds of auscultation and percussion, or the judgment uninformed by experience may mistake the conclusions which ought to be deduced from

B

them, if correctly interpreted; but undoubtedly the opinion of a number of accomplished physicians would generally be unanimous in deciding on the actual lesion in any case submitted to them. Discordance of opinion would only come out when the question arose as to how such a condition had come about, and how far it might be amenable to treatment or compatible with life. I do not, therefore, propose to inquire what is the significance of the sounds of the heart as heard in disease, but assuming that my readers have made themselves familiar with all that the stethoscope can teach of alteration in sound and rhythm as indicating corresponding changes in function and structure, I propose to inquire how far such changes can be traced to gout and rheumatism as their cause.

No subject has more engaged the speculative talent of medical writers than the causation of these disorders, and the best means for their alleviation. And though certain facts have been proved concerning them which were previously only suspected, yet it seems that we are still nearly as far as ever from a thorough understanding of the point of departure from health, which leads to the series of symptoms to which the names so familiar to our ears have been respectively applied.

To Dr. Garrod we owe the certainty of the existence of an excess of uric acid in the blood during the gouty paroxysm, and its absence in rheumatism. Many had groped in the dark at the expression of this fact before, because of the manifest relation that exists between the presence of an abnormal amount of uric acid in the secretion of the kidney, and the tendency to gout. We are now in possession of the knowledge that the uric acid is circulating in the system ; we know, too, that it gets fixed in the inflamed joint, but nearly all that associates these two facts with the attack of gout is purely hypothetical. It is far too mechanical a view to suppose that gout is nothing more than an accidental excess of uric acid in the blood : that the inflammation of the joint is simply the effect of the same acid in the form of urate of soda exuding through the vessels, and irritating the ligamentous and synovial structures. The fallacy of such reasoning becomes more apparent when we find Dr. Garrod asserting " that true gouty inflammation is *always* accompanied with a deposi-

tion of urate of soda in the inflamed part."[*]
So far as it is possible to trace the reasoning
by which such a conclusion is arrived at, it
might be stated as follows:—Gout is essentially
known to us by the inflammation of the joints,
and this consequently is true gouty inflammation. Any joint so attacked has been found
on examination to bear traces more or less
distinct of the deposit of urate of soda. Hence
all true gouty inflammation is accompanied
by this deposit. I need only appeal to Dr.
Garrod's own logical faculty to admit the
fallacy of the argument so stated, and I do
not find that he has anywhere brought forward
further proof of his assertion.

It is indeed quite true that in the case of
the joints we find the existence of gout, the
local inflammation, and the deposit, all har-
monising together; but does it necessarily
follow that if, during the existence of gout,
inflammation of any tissue does not present
the same deposit, it must be excluded from our
idea of the disease? Must we of necessity
find urate of soda in the stomach and the
bronchi before we can admit gouty gastritis or

* 'The Nature and Treatment of Gout,' by A. B. Garrod, M.D.,
p. 340.

gouty bronchitis? Must we again assume that as "the deposited urate of soda may be looked upon as the cause, and not the effect of the gouty inflammation," no gouty inflammation can occur till such a deposit has taken place and that it must have happened wherever it exists? Such assumptions make large demands on our belief, and to me it appears that the very fact that urate of soda is not found in parts such as the bronchi, where an inflamed condition of the membrane is so often associated with gout, is of itself a proof that "true gouty inflammation" is not always associated with or caused by the deposit. This conclusion acquires additional force from the consideration, that though the deposit and the inflammation are associated together in the joints, the urate of soda is seen in other parts without any evidence of its exciting inflammation there.

There are other circumstances, too, which seem to point out that there is something more than a mere chemical agency at work. In the first place, the hereditary tendency so strongly marked in certain individuals is one of the features of the disorder which has been long known, and has been more universally accepted with reference to this than to almost any other malady. Then, again, the kidneys in a state of health can and do secrete such an amount of uric acid that, however quickly it is generated, no accumulation takes place in the system under ordinary circumstances. In gout, without any evidence of disease, their function becomes so far suspended that the excess remains during the whole period of the attack. Lastly, in the process of cure, how often may the disorder be suddenly cut short by having recourse to certain remedies which act as specifics, while others that would seem exactly suited to remove the excess of uric acid from the blood are quite powerless to produce this effect? Each of these points demands careful consideration.

No theory can for a moment pretend to explain the mystery of hereditary transmission. It must ever remain inexplicable to human reason why not only the peculiarities of species and race, but even those of feature and character, are stamped on the germ-cell *ab initio*, and are gradually developed at different stages of its existence. No less must the transmission of actual or latent disease remain unexplained. We must be content with sifting the evidence of its reality, and the laws of its manifestation. These differ in each of what are commonly known as hereditary disorders.

The evidence of a syphilitic taint may be found in the fœtus *in utero*, and is always developed in the early period of extra-uterine life. The strumous diathesis may either show itself in childhood or may lie dormant till after puberty, the particular age at which it appears varying in different families, but comparatively few escaping when the tendency is strongly marked. Pulmonary tubercle is most notably a disease of early adolescence, and though it may number its victims among almost all ages, its range as a hereditary disease is comparatively small, and it does not appear that the age is much influenced by the intensity of the predisposition. One only of a family may die before the age of twenty while the rest escape; or the whole may be cut off at nearly the same age though none die very young.

The appearance of gout follows laws somewhat different. It is, perhaps, never seen actually in childhood :* its first appearance is

* Dr. Gairdner refers to only a few cases among young persons

usually rather late in life; acquired gout,
indeed, may be said never to come on before
the middle period; and hereditary gout, when
appearing sooner, does so apparently in proportion to the strength of the tendency which has
been handed down. If all the sons of a family
suffer, some among them are almost certain to
have it very early. The exemption of females
is no less remarkable, and unquestionably
points to the necessity for the operation of
some exciting cause before the hereditary
tendency can assume the form of actual disease.
In no case is there at all the same probability
that every member of a family will suffer, as
in consumption ; and perhaps total abstinence
from alcoholic stimulants, and a sparing diet,
so far as animal food is concerned, might, if
carried out from an early age, secure an individual from the hereditary disorder, however
strongly it was impressed on his organisation.
Moderation, indeed, we know to be successful
in a large number of cases, and it is not improbable that whenever gout appears, whether
the tendency existed or not, the individual has
at some time or other exceeded what was for

in his very extensive experience, and cites an instance of a child
of eleven as quite exceptional.—'Gairdner on Gout,' p. 269.

him the bound of moderation. No arbitrary standard can be fixed which shall be the same for all. To one, alcoholic stimulants seem in the present condition of the species a necessity for the maintenance of health; to another, they are a luxury which in certain quantity do more good than harm, and greatly add to the happiness and wellbeing of the individual; while, to a third, even small quantities do harm, and, if habitually employed, tend to produce permanent mischief, either by causing the degeneration of some internal organs, or by developing gout or delirium tremens. General experience seems to point out that all stimulants do not act alike in these respects, but that the after consequences of indulgence vary with the kind of stimulant which has been employed; but the rule is by no means invariable, and gout, which is supposed to be especially the inheritance of the rich and the fruits of port wine, may be acquired by the poor from excesses in gin and beer.

Indications of this nature, while they serve to show that there is something needed to make up the total of history, and symptoms which we call gout beyond the mere excess of uric acid, seem to go further and suggest that

the first change must be in the molecular structure of the blood itself. It has been shown
that even to the rude tests of chemistry a
change is wrought in the constitution of the
blood by long habits of intoxication, that the
tissue metamorphosis is retarded, and the effete
particles tend to accumulate in the circulation.
The nervous symptoms of delirium tremens are
much more probably dependent, like those of
fever, on blood changes than on structural
alteration of the brain; indeed, the analogy
between them and those of typhus has always
appeared to me very striking. That gout in
the same way depends upon blood change,
perhaps no one will venture to dispute; but
those who hold to the chemical theory, view
the change as merely consisting in the accumulation of uric acid in the serum, a theory which
seems inadequate to explain the whole of the
phenomena.

The chemical theory fails more evidently
when we come to consider the relation of the
function of the kidney to the gouty attack, and
points directly to some anterior agency as the
cause of the disorder. Inexplicable as the laws
of secretion may seem, there is unquestionably
some relation between the organ and the sub-

stance which is eliminated through its agency. The liver will not secrete uric acid, the kidney never produces pepsine; and while there are some secretions which contain principles of paramount necessity to the economy, saliva, pepsine, &c., others are mainly, if not entirely, made up of useless material. Such is the kidney secretion, and it is very remarkable that of those which are subsequently employed for the purposes of nutrition, some, at least, of the component principles are formed by the secernent organ itself, while the excrementitious secretions seem to consist chiefly of effete particles already formed in some other part of the body, which are seized and separated from it by the organ appointed for their elimination.

In their results the effect of the suspension of the function in either case is widely different. There is, on the one hand, an arrest of the development of some principle more or less essential to life, and a consequent interruption of the general process of nutrition; and, on the other, an accumulation in the blood of some element which ought to have been got rid of, and a consequent condition of impurity which renders it more or less unfit to maintain healthy vital action. In the one, the disease somewhat

resembles starvation; in the other, an actual
condition of blood poisoning is produced. Such
a contrast is afforded, for example, by any
interference with the secretion of the stomach
as compared with the suspension of the function
of the kidney. When the gastric secretion is
deficient or disordered, so that a proper amount
of pepsine is not formed, there is no accumulation of that element in the blood, but the
health fails because the albuminoid principles
are not perfectly digested and converted into
healthy blood. But when the secretion of the
kidney is imperfectly performed, the urea, the
lithates, and the phosphates which are its chief
ingredients accumulate in the blood, and the
urea especially acts as a direct poison on the
nervous system.

When this knowledge is applied to the part
played by the kidney in gout, it must strike
every one that there is an essential difference
between that suspension of its action which
puts a stop to the secretion of all the elements
alike, and the special interruption which occurs
in gout when the uric acid is mainly deficient
and accumulates in the blood. The time has
not yet arrived for a full explanation of this
peculiarity. Gout very generally lurks in the

system for a long period before it makes its true character known by the attack on the joints, and is perhaps most familiar to practitioners under the form of gouty dyspepsia. Ordinary remedies are tried and fail because there is something behind and beyond the dyspeptic symptoms which they do not touch. Occasionally by the judicious regulation of diet, the stomach recovers its tone, and the patient may go on for some months or years longer before the gout comes out, but ultimately a regular attack establishes the correctness of the diagnosis if such be the cause of the phenomena observed.

One can hardly refrain from speculating on such a subject, though knowing how vain such speculations are. We naturally ask ourselves, why, when different persons are severally indulging too freely in alcoholic stimulants, the symptoms are so different in each. To some it occasions constant headache, and perhaps they are the most fortunate, because they are constantly reminded by their own discomfort of the folly of indulgence ; in others it produces biliary derangement, which is probably at first only functional, but ultimately leads to serious disease of the liver. Sometimes the

organ presents the character of cirrhosis,
sometimes it becomes fatty or lardaceous. In
the former case the progress of the affection
is so insidious that long after the habit of
indulgence has been abandoned the patient has
to pay the penalty of his folly in a lingering
illness which gave no indication of its advent
till gradually increasing enlargement of the
abdomen, with the symptoms of dropsy, prove
that the blood-current is arrested in the
shrunken liver. In the latter, it is not difficult to recognise from the general unhealthy
deposit of fat, and the prominent yellowish
eye, a tendency which will one day or other
end in a degeneration of that and probably
also of other internal organs. Such is the
aspect of the habitual sot; and it is worthy of
note that he is peculiarly liable to attacks of
delirium tremens.

On the other hand, the "horrors" may be
the first indication of mischief in a man who
is not constantly tippling but indulges every
now and then in a drunken debauch. I think
it probable that when this occurs, the consumption of spirits has brought on an unusual
degree of anorexia, has indeed impaired the
powers of digestion, and destroyed the secret-

ing power of the stomach, so that the early appearance of the nervous symptoms may be explained by the absence of proper nutriment coinciding with the abuse of alcoholic stimulants which have almost entirely taken its place.

Then, again, we have as a consequence of excessive stimulation the most insidious, perhaps, of all the diseases resulting from such habits—degeneration of the kidney with albuminous urine. The victim of this disease has nothing to remind him that he is doing himself permanent injury by his folly : he is willing to compound for a little passing discomfort in order that he may continue his so-called pleasure, but nothing gives any hint of the gradual disorganisation of his kidneys which is going on.

Each of these forms of disease may be combined with gout, but it may also be quite independent of all. We are not indeed surprised to find that it is very often associated with disease of the kidney. Not only is an abnormal condition of the secretion constant during the gouty paroxysm, but we should naturally expect that when the excretion of all the elements of the urine is interfered with, more trivial circumstances than usual would give rise to the retention of an excess of uric acid. It is more particularly with one form of degeneration that this association is traced, but as gout may occur without the kidney complication, so, too, in numberless instances, does the very same form of albuminuria exist without any indication of gout; and that, too, it may be remarked, in the very cases in which it might be looked for, viz., when the disease has been brought on by excess.

Closely allied as all these affections are in their origin, it is not difficult to trace out some broad distinctions in their causation, which, though imperfectly defined, are acknowledged as the results of practical observation and experience. Gout, it may be said, is more distinctly brought on by the combination of good living with stimulants, such as is usually the case with port wine drinkers: and it is more likely to be developed by wine, and especially red full-bodied wine, than by spirits or ales. A hobnail liver is more certainly the result of spirit-drinking, and albuminous urine of a large consumption of beer. Delirium tremens, on the other hand, is unquestionably the expression of want of food along with excess of stimulant. All this may be very true, but it does not really give us clearer ideas of their mode of production; and marked exceptions to any such general rules constantly occur, especially when, as is not unfrequently the case, two sets of symptoms may be traced in the same individual.

It seems impossible to doubt that the first departure from health, or at least from a condition compatible with health, must be found in an altered condition of the blood. It is to be remembered that there are several series of changes going on in that fluid from day to day, some of which are quite transitory, while others are more or less permanent in character. First we have changes in the condition of the serum, which are in progress every hour; it contains more or less water, saline substances, nutritive material in a state of solution, and effete matter. In a very few hours a large quantity of water may be got rid of by violent perspiration, excessive kidney secretion, or bowel flux. A very few minutes serves for the transmission of a saline substance from the stomach to the kidney, where it may be traced in the urine; and if taken in larger quantity, two or three hours will suffice for

many salts to pass off by the bowels. Excrementitious particles are more slowly got rid of
by the various secretions, and the nutritive elements probably remain a very much longer
time before they are converted into tissue of
any kind; and no doubt a considerable amount
of this material never gets any further than
the blood, but is used up there for other purposes—that of respiration, for example—without ever being actually and directly applied to
nutrition. Another series of changes is connected with the formation, alteration, and disintegration of blood-globules and colourless
corpuscles. That these are sometimes very
rapid in their progress I see no reason to
doubt, but still they must be regarded as being
more stable and permanent in their character
than the changes which go on in the blood-
serum. Now it would appear that the various
stimulants taken up into the blood and passing
round in the circulation act in one man more
injuriously on the brain, in another on the
liver, in another on the kidney, and as a consequence of this injurious action being long
continued and often repeated, serious, or even
irreparable organic change at length takes
place. In the very same way I think it must

sometimes happen that the blood globules themselves are injuriously affected, as, for example, by the repeated introduction of the gout-producing elements into the circulation.

I confess that there are many difficulties attending such an hypothesis, and one not the most easy to explain is the permanent character of the gouty tendency once developed. It seems hardly credible that a fluid changing its condition so constantly, that globules which at best must be very short-lived, should be the means of maintaining such a tendency for life. But we are not without analogies which render such an explanation possible. A flesh wound or a burn leaves a cicatrix which is formed out of plastic material, and is not of the same structure as the parts which it unites. To the end of life it maintains its distinctive character, and the formative cells, upon which growth is supposed to depend, take on a different mode of development in the cicatrix from that which they would follow were they subject to the ordinary laws of growth, different, too, from that of the cells which absolutely touch them. It certainly is not the case that a cicatrix once formed remains unchanged, and that development and decay in it are alike unknown. What,

for example, can be more distinct than the
growth of the vaccine scar, so small on the
infant, so much larger on the man? If for a
moment we consider the nature of those cells
out of which its growth and development come,
we may well ask wherein they differ in their
permanence from blood-cells ; and it is remarkable that while a scar has been formed which
lasts a lifetime, some change has also taken
place in the system at large with reference to
its susceptibility to the influence of what every
one regards as a blood-poison. In other blood
diseases similar changes are seen, and all present a certain degree of permanence. To what
else can we look but to the blood-globules to
account for such alterations ? To them must
be due the degree of permanence which they
exhibit. In fact, just as each disintegrated
portion of structure is replaced by another
bearing a close resemblance to itself, in such a
manner that under all his changes the man
remains the same individual to the end of his
life, so the blood-globules, having received a
certain impress, must be succeeded when they
become worn out, by others which have a
general resemblance to them. In course of
years the features change, so that a brother

may not be recognised after a long sepa-
ration, especially if a severe and exhausting
illness have occurred during the interval, and
much disintegration of tissue have taken place ;
the new cells are *nearly* but not *quite* like those
which they replace ; the man is only like his
former self. So a blood-change may gradually
wear out in some individuals, by the successive
crops of blood-globules being not exactly like
those that preceded them.

The means of confirming or contradicting
such an hypothesis are as yet wanting. We
cannot pretend to exhibit under the microscope, or by chemical reagents the change produced by vaccination, or smallpox, or scarlet
fever, which protects the individual more or
less during life from the consequences which
follow the reception into the body of certain ex-
halations : neither can we trace the strumous
or the cancerous tendency, whether located in
the blood or in the solid structures of the
body. In such circumstances the physical
search for the gouty tendency must be alike
unproductive, and speculation can alone suggest a theory which may serve to explain the
phenomena. I trust that I may be excused
for having ventured on such a subject, but it is

one which I know has occupied the thoughts of enquiring minds, and I have particularly heard my friend Dr. Gairdner, who is so great an authority on gout, express the opinion as the result of his thoughtful investigations, that change in the blood-globules was the foundation of the disease.

Such speculations cease to be profitless when we turn to the treatment of gout. Were it true that non-elimination of uric acid from the blood is the sole cause of gout, it is manifest that we have only to secure complete alkalescence of the blood and the urine—a condition in which deposit of uric acid is *chemically* impossible—to save our patients from any of the evil consequences of its presence. According to the chemical theory, gouty inflammation is only inflammation with uric acid deposits, and therefore chemically the conclusion is unavoidable that if the deposit be impossible, gouty inflammation is impossible also. A man with a gouty tendency has only to drink enough potash water or lithia water and he need never have the gout, port wine and beefsteaks notwithstanding. In our view of the subject the retention of the uric acid is a symptom, a consequence of the attack of gout and not its cause.

The good living and the stimulants do not simply cause an excess of uric acid to be formed, but they end by causing some more permanent change, and probably one affecting the blood-globules; which reacts on the kidney putting a stop to the excretion of uric acid, and causing its retention in the serum, where, passing in the round of the circulation, it is very apt to become deposited as urate of soda. Experience would seem to lend its sanction to the latter view. No amount of artificial alkalescence of the blood and urine will put a stop to the paroxysm when it has once commenced. In old standing cases the subacute inflammatory condition will last for months in spite of alkaline treatment fully and fairly carried out. No one who has compared in an early and acute attack the power of alkaline remedies with that of the so-called specifics, such as colchicum, in arresting its progress, can doubt that there is a disease to which the name of gout is applied distinct from the excess of uric acid in the blood serum which attends its progress.

It is quite true that a due supply of alkali is of great value both as a prophylactic and a remedy in gout; and that a patient to whom it is properly administered will escape many of the worse evils which follow in the train of the disorder. It is perhaps also true that a gouty attack is more safely, if not so expeditiously, passed through without any aid from colchicum. But the very shortest experience in treatment proves beyond all question, that when colchicum is first administered to a gouty patient, its action is something perfectly different from the effect of the administration of alkalies. By the one remedy the disease appears to be almost immediately arrested, the existing inflammation rapidly subsides, no fresh joint is attacked, and in a few days the patient is well. The other mode of treatment seems only to modify the severity of the seizure without arresting its progress—the inflammation of the joint continues for some time in a modified form, fresh joints even may be attacked after the remedy has been freely administered.

In subsequent attacks the difference between them is not so marked, and the impression rather gains ground among medical practitioners that the colchicum treatment, if more rapid and more certain in the first instance, may, if used in excess, be ultimately injurious to the patient. At least it may be said with great confidence

that its management requires skill and judgment, and there is less risk of harm from injudicious treatment when the remedy employed
is one calculated only to avert the evil consequences arising from the presence of an excess
of uric acid in the blood, and not to arrest
the attack; nature being left to throw off by
her own process the accumulation which has
taken place, and to relieve the system of the
condition of blood-poisoning brought about
by it.

CHAPTER II.

———◦◦◦———

Distinction between Rheumatism and Gout — Rheumatic Gout — Its progress — Disorganisation of Joints — Dr. Adams — Osteo-Arthritis — Acute Rheumatism — Muscular — Chronic — Active and passive motion — The acid of Rheumatism — Of Dyspepsia — Rheumatic inflammation of the heart — Discovery of the relation — Dundas — Bouillaud, &c. — Proportion of cases—Statistics of St. George's Hospital — Subacute Rheumatism —Rarity of Cardiac Inflammation in this disorder.

RHEUMATISM presents many features which possess a remarkable similarity to the symptoms of gout, and yet the two diseases in reality stand wholly apart in their origin and progress as well as in their results. Few points in pathology are more remarkable than the difference between the forms of cardiac disease which usually result from either. The points of resemblance are so numerous that it is somewhat difficult to comprehend why in this respect they should be so dissimilar. It is true that between a first attack of acute rheumatism and acute gout the line of demarcation is pretty distinctly drawn, and there is no great chance of an intelligent practitioner making a mistake in diagnosis. But when the disease recurs, or presents itself in a subacute or chronic form, it may require considerable scientific skill and practical experience to determine under which head the case is to be classed.

The term rheumatic gout has been used by systematic writers to include a form of disease in which the symptoms were not distinctly either rheumatic or gouty. The name is an unfortunate one, because while the condition intended to be expressed is one sufficiently recognised and perfectly distinct from either disease, the name has been not unfrequently employed merely to cover the ignorance of the person using it, and save him from the necessity of expressing in so many words whether the patient were suffering from the one disease or the other. It is therefore much to be desired that rheumatic gout should be wholly banished from medical nomenclature. It implies a theory which is known to be false, and it affords an opportunity to those who do

not care about accurate diagnosis to escape
from the necessity of expressing an opinion.

Dr. Garrod has proved that the cases so
denominated present none of the real features
of gout, inasmuch as an excessive formation of
lithic acid is wholly wanting. To Dr. Adams
we are indebted for a most careful analysis
of the special condition of the articular surfaces in this disorder, and from his researches
it would appear that the changes which take
place in the intimate structure of the joint are
of a character wholly different from anything
found in true rheumatic affections, from which
it must be as completely separated as from
gouty disorders. His name of osteo-arthritis
will perhaps be accepted by the profession as
an improvement on the former appellation
though rather a clumsy and unmanageable
one. We are still left in the dark, however,
as to its mode of causation and its relation
to true rheumatic affections.

Commencing very often as a mild form of
rheumatic fever, it probably excites very little
apprehension at its first appearance. No
cardiac complication exhibits itself, the fever
is very moderate in intensity, and the number
of joints attacked simultaneously not by any

means great. It is naturally assumed that such a condition will speedily yield to treatment, and it is not till weeks have passed without any material change in the symptoms of the case that its real nature begins to be suspected. Joint after joint is attacked by the malady, and while the acuteness of the pain accompanying it subsides after a time, it is found that the slightest movement is attended with suffering, and the patient is seen gradually sinking into a condition of utter helplessness, as each joint in succession becomes permanently disabled. Such are the characters of a disease which not unfrequently commences in females at or near the time of the establishment or cessation of the menstrual function. For very many months, or more probably for years, a certain degree of inflammatory action is maintained, and the motion of the joint which is more or less restricted by the thickening of the surrounding tissues is still further limited by the pain attending its movement; ultimately considerable distortion supervenes, and all the joints are more or less stiffened into some unnatural form. When movement of any sort is retained it is much interfered with by the destruction of cartilage, and though comparatively free from suffering, the patient remains a hopeless cripple for life.

Closely resembling a subacute form of rheumatism in its commencement, while in its advanced stages the joints of the hands and feet bore in their distortion some resemblance to the disfigurement of gouty deposits, it was not unnatural that the name of rheumatic gout should have been employed to describe them. But Dr. Adams shows that this is not the only form in which the affection exists. In single joints eburnation of the cartilage may result from its presence, and render them more or less useless for the purposes of motion. I must refer my readers to his able descriptions for fuller information on the subject.

In speaking of disease of the heart as of rheumatic origin, it must be remembered that it is to acute rheumatism almost alone that such a result can be ascribed. It is highly probable that at some future period a more distinct relationship may be traced between the acute and some of the chronic varieties; and it is not improbable that some of them may be as completely separated from rheumatism as osteo-arthritis now is. There are, to all appearance, cases of chronic rheumatism which have no relation whatever to the acute affection, while there are others that merge into it, either by springing from it in the first instance, or by assuming during their continuance symptoms of a decidedly acute, or at least subacute character, not unfrequently indeed terminating in this manner. That these are really different affections, though included under one term, seems to offer an explanation of many anomalies, but so long as there are no circumstances by which they can be distinguished beyond the fact that, in some cases, acute symptoms have at one time appeared, we should not be justified in attempting to sever them.

The pathology of rheumatism is still quite as obscure as that of gout. Indeed, it may fairly be alleged that we know less of its causation and are more uncertain of the manner in which its symptoms are brought about. That it is somehow connected with an excess of acid in the system no one can doubt who has had the least experience of the disease. The perspiration is in the acute disease intensely acid and sour-smelling. The urine is acid, and owes to this its constant turbidity and its copious deposit of acid lithates. But how all this originates is very unintelligible.

It is very often assumed that rheumatism is brought on by exposure to cold; and the explanation of this commonly received opinion is probably to be found in the circumstance that a stiff neck or some other form of muscular rheumatism can be so constantly traced to a draught of cold air. But we do not find that muscular rheumatism is the commencement of that more formidable disease known as acute rheumatism or rheumatic fever. All experience goes to prove that though called by the same name the two affections are very distinct from each other. Perhaps to this circumstance may be attributed in part at least the confusion which exists in men's minds as to the essential character of the disease. No theory can be admitted, which does not equally explain the symptoms depending on rheumatic fever, on muscular and chronic rheumatism, on that which specially affects the ligamentous structures, and that which produces an excessive secretion from the synovial membrane.

If the subject were studied aright there ought to be a clear distinction between cases so totally dissimilar. The result of my own observation specially directed to this point, has been a firm conviction that cases of chronic rheumatism are not as a rule, more constantly associated with a liability to the acute form than any other set of cases that may be selected. And if it be doubted whether the testimony of patients on such a point is to be relied on, I can at least affirm with perfect confidence that they did not, so far as I have observed, present more instances of disease of the heart, which as we shall find is so constant an accompaniment of the acute disorder.

It is, however, a matter of extreme difficulty to draw the line between the various forms of joint disease when there is no suppuration, no ulceration of cartilage. The objective symptoms consist only of swelling of the joint, redness and heat when present, and the diagnosis rests mainly on the subjective phenomenon of pain complained of by the patient. It is true indeed that the extreme tenderness of an acute attack can never be simulated; but when there is no swelling, no heat or redness, the indication derived from mere sensation is not wholly trustworthy. Aches and pains are constantly regarded as rheumatic which may be due to totally different causes; and while one person seems to suffer extremely, though there be no external evidence of disease, another makes light of pains which cannot be supposed to be trivial if the manifest changes in the part affected be taken into consideration.

For practical purposes it seems at present to be the safest course to regard the series of cases which bear this common designation as divisible into three principal varieties which have really distinct causes and consequences. 1, Acute rheumatism; 2, Muscular rheumatism; 3, Chronic rheumatism. Under the head of acute rheumatism we must place a minor section of subacute cases in which the symptoms are quite analogous though less severe, and which merge into the acute variety by such insensible degrees that it is impossible to draw the line between them. On the other hand inflammation of the joint tending to suppuration must be excluded as owning a totally different origin, as well as those cases in which the rheumatic affection is dependent on gonorrhœa. Here too we meet with that form of rheumatism to which the name Synovial has been given; and while they present characters which might fairly be called subacute, they differ from ordinary rheumatism in being accompanied by an excessive secretion of synovial fluid, without much affection of the ligamentous structures. This is no unnecessary refinement when the course and progress of such cases is considered.

The symptoms of muscular rheumatism are generally not hard to trace, and are only liable to be confounded with neuralgia. But in chronic rheumatism diagnosis is sometimes more difficult, as it is necessary to exclude not only cases dependent on a specific poison such as the syphilitic, but also, in my view of the disorder, those which are really the sequel of an acute or subacute attack, and have not begun as so many do, simply as a chronic disease from the first. Sometimes there has gradually supervened a change in the appearance and mobility of the joints which is quite characteristic; in other cases there is nothing perhaps to indicate any departure from health, except that any movement whatever is a cause of suffering.

To one point in the diagnosis it seems to me that sufficient attention has not been paid—the distinction between pain caused by active and passive motion. A moment's consideration must prove the importance of discriminating them. If pain be only felt when a voluntary effort to produce movement of the limb is made, and the joint can be moved by the hand of the observer without suffering, it is plain that the tenderness is not in the joint itself, but in the muscular apparatus by which it is moved. If, on the contrary, passive movement causes pain when the volition of the sufferer is not called into exercise, it is equally manifest that the structures of the joint itself—ligament, carti-lage, or synovial membrane—must be the seat of pain. Simple as the observation may seem, the results are not always consistent, because it is not always possible for the patient to be wholly passive in the hands of his atten-dant. The fear of his causing pain by moving the limb very often puts all the muscles on the stretch, and no movement can be made without pulling against some one or other of them, in such a manner that when muscular rheumatism is present, the one affected is almost certain to suffer.

In rheumatic fever we recognise a condition which in its typical form cannot really be simulated by any other disease. It presents two distinct series of phenomena, the general pyrexia and the local affection. Sometimes the one series of symptoms, sometimes the other takes precedence in order of time, but they are closely linked together, and it is quite clear that the one is just as dependent on the all-pervading cause as the other. The fever is no more produced by the inflammation of the joints than that inflammation is excited by the fever; both acknowledge a common origin. And whatever that common cause may be it is invariably attended by excessive acidity. It is a mere juggle of words to define it as mal-assimilation. When such a term is employed it only means that it does not owe its origin to any specific external cause, but that it springs from within; it gives us no insight into the manner of its development.

The existence of some influence of a general character, associated with, and probably originating the disease, is proved by the presence of acid in all the secretions. Tracing it backwards to its origin we shall not be far wrong in asserting that this acidity begins in the stomach. The gastric juice is itself acid, and loses much of its solvent power if it be rendered neutral or alkaline. But the acidity may be much in excess of the normal

standard, and the food introduced into the
stomach may there undergo a process of fermentation in which acid is further developed.
Most of my readers must be familiar with acid
dyspepsia and heartburn; but the question at
once arises, does this lead to rheumatism? I
am convinced that it does not. I shall never
forget a most remarkable instance of this
disorder of stomach, which came under my
notice some years ago. It was at the time
when the attention of the profession had been
called to the existence of a peculiar fungoid
growth to which the name of *Sarcina Ventriculi*
was given. When present in the stomach this
little parasitic vegetable occasions the most
rapid fermentation of the alimentary substances,
and the most extraordinary development of acid
in the stomach. This acidity was carried to
its highest pitch in the patient referred to, and
it was most remarkable that at the period of
digestion, at which the acid formation was most
rapid, the kidney secretion was so very alkaline that the phosphate of lime crystallised into
prisms of more than an inch in length. If
vomiting occurred, as it often did soon after
this, the urine retained its alkalescence for many
hours; if the food was retained and passed

into the bowels, it very soon became remarkably acid. In this case, there was no tendency to rheumatism whatever, and the teaching of such an observation is, I think, subversive of the doctrine that the acidity of rheumatism is due simply to the stomach. There is some peculiar tendency in the system to develope and retain the acid which does not exist in mere acid dyspepsia; and if the acidity show itself in the process of digestion, it has some more deep-seated cause.

It is very generally alleged that the acid of acute rheumatism forms the link between the articular and cardiac inflammations; but we are not yet in a condition to prove that the acid is the cause of the joint disease, and it would therefore seem rash to speculate as to the causation of this further complication. The one is an invariable product of the disease, while the other is only occasional. Its history has been recently pretty fully explored; and it may be admitted that almost every form of heart disease, as seen on *post-mortem* examination, may be traced back to inflammation during an attack of rheumatic fever. In some cases, of course, it is plain that the change in structure is not inflammatory, and, therefore, that it acknowledges some other origin. But there is scarcely one lesion of any importance which does not find a parallel in some case of rheumatic inflammation. This is true as well of the thickenings and fibrous or fibroid deposits, as of the degenerations, whether fatty, athero-matous or calcareous.

In the first instance, the change is one which belongs to the larger group of what are called inflammations; a change primarily affecting the solid structures of the part producing a local stasis of blood or congestion of vessels, and followed by exudation of the water and soluble constituents of the blood. To this exudation is due the thickening, œdema, and fibrinous deposit. At some parts of the endo-cardium, it may be considered doubtful whether the lymph found on its surface has not come directly from the mass of the circulating fluid constantly passing over it, rather than from any process of effusion from the membrane itself. This question is, after all, of very secondary importance, because the adhesion of the deposit to the naturally polished surface, proves that change of some sort has taken place in the lining membrane of the endocardium ; and in the pericardium the source of the effusion cannot be doubtful. We are bound to admit that whatever may ultimately be proved regarding the mode in which the lymph-beads on the valves of the heart are formed, there can be no question that both within and without the heart, the serous covering tends to undergo changes of a kind which are called inflammatory, during the existence of rheumatic fever.

The subject was very carefully investigated by Bouillaud, whose merits as an observer would have been more appreciated but for his method of treatment, which subsequent observations have proved to be not only useless, but actually injurious. It is interesting now to look back and trace the steps by which we have arrived at our present position of comparative familiarity with forms of disease which were quite unknown at no very remote period. In a paper dated 1806, we find Mr. Dundas, Serjeant-Surgeon to the King, calling the attention of the celebrities who formed the Medical and Chirurgical Society of that day, to what he simply styles, "a peculiar disease of the heart." Nine cases of this disease having come under his care in thirty-six years, he thinks that it is not very uncommon. His attention seems to have been first arrested by

a case which was examined by himself after
death so long ago as 1770. in which pericarditis
had followed on acute rheumatism: and he
states that "in all the cases which he had seen,
the disease had succeeded one or more attacks
of rheumatic fever." The majority of the cases
recorded were fatal, consisting as they did only
of the most severe examples; cases in which
no physical sign was available for diagnosis,
beyond that of heaving impulse, rapid and
occasionally irregular pulse — when friction
sound was unknown, and it was considered to
be a sign of very serious disease that "the action of the heart could be distinctly heard."
The character of the disease was only obscurely
recognised by the symptoms of anxiety and
oppression at the præcordia, with extreme dys-
pnœa and apprehension of death. It was not
possible to trace the various steps by which the
disorder advanced; but the fact then first became generally known to medical men that
such a disease might occur as a sequel to rheumatic fever.

These observations led to others in this
country, but seem to have been comparatively
unknown on the Continent. Laennec failed
to trace the auscultatory phenomena of cardiac

inflammation chiefly because he knew not when to look for it, and the chances of making observations enjoyed by him were so few. His discovery of the employment of the stethoscope for the examination of the thoracic viscera very soon, however, bore fruit, and the elaborate memoirs of Bouillaud in France, and the labours of Hope and others in this country, speedily familiarised the minds of medical men with that which is now recognised as one of the most distinct instances of causation, with which we are acquainted in the history of disease.

Attempts have been made from time to time to calculate the proportion in which cases of rheumatic fever were complicated with one or other form of cardiac inflammation. We can scarcely trust the statistics of Bouillaud, because he takes no account of the possibility of a *bruit* being caused by an anæmic condition which was not unlikely to arise from his excessive depletion. He gives us the very high rate of eighty-six per cent. as the proportion of cases in which the heart was affected among those which came under his care suffering from acute rheumatism.

The result of my own observations in St.

George's Hospital nearly fifteen years ago[*] gave a percentage of forty-four in which the heart was either attacked by inflammation at the time of observation, or had suffered in a previous attack of rheumatic fever.

Dr. Fuller,[†] in tabulating the cases of a few previous years, obtained a considerably higher percentage in the same hospital, while Dr. Dickinson,[‡] writing in 1862, finds the proportion of heart affections very greatly reduced since the employment of the alkaline treatment.

These facts seem to point to considerable improvement in methods of treatment, and lead us to hope that ultimately, the numbers suffering through life from rheumatic affections of the heart will be far fewer than they are at the present day.

The association of inflammation of the heart with subacute rheumatism is certainly a rare event, except in childhood. Some amount of inflammatory action, and of febrile disturbance, would seem to be essential to its development. Subacute rheumatism is generally only rheu-

* 'Med. Chir. Trans.,' vol. xxxv.
† 'On Rheumatic Gout and Sciatica,' by H. W. Fuller, M.D.
‡ 'Med. Chir. Trans.,' vol. xlv.

matic fever deficient in these particulars, and
therefore we are not surprised to find that the
risk of cardiac inflammation is so small. Practically it is very difficult to specify the condition to which the term subacute is really
applicable.   In childhood the joint affection
may be slight while the febrile disturbance is
very considerable; and perhaps the evidence
derived from general symptoms is more
useful as a guide in attempting to discriminate the two sets of cases than any
other.   With reference to the chance of
the heart being implicated, it may be said
that in childhood any rheumatic attack pre-
senting febrile symptoms is likely to be accompanied by inflammation of the heart; that
at the age of puberty a distinction can readily
be drawn between cases in which such danger
is imminent, and cases in which it is not
probable; and that after adult age is complete, a very marked febrile and inflammatory
character must accompany the disorder in any
case, before there can be the smallest probability of cardiac inflammation.

Constantly, however, at all ages, and under
all circumstances, previously existing disease
may be discovered during an attack of sub-

acute rheumatism. It may be impossible to decide that it has not begun during the present illness from any signs actually present; but if there be a history of a previous attack of greater severity, if there be now comparatively little fever present, it is reasonable to conclude that the disease of the heart is not recent. This conclusion is still more unquestionable in chronic rheumatism where every murmur must be dealt with as having nothing whatever to do with the present condition of the patient. Whatever may be the relation of chronic to acute rheumatism, it certainly has no influence whatever in developing cardiac inflammation. It is not impossible that the very same circumstances which have produced the rheumatic tendency have also led to degeneration in some of the internal organs, and thus the chronic diseases of the heart may be linked with chronic rheumatism, but the one can never be regarded as the cause of the other.

CHAPTER III.

———◦———

The earliest indication — Altered rhythm — Its relation to "bruits" — Discrimination of abnormal sounds — Friction — Its character and rhythm — Endocardial Murmurs — Their causes — Regurgitation — Discrimination of Anæmic Murmur — Not produced by the Valves — Its situation — Its occurrence in Acute Rheumatism — Valvular insufficiency without disease of Valve — Progress of Inflammatory Disease — Consequent enlargement of Heart.

THE earliest intimation we have of inflammation attacking the heart or pericadium in cases of acute rheumatism, comes in the form of altered rhythm of the heart's sounds. This generally precedes the development of any abnormal sound whatever. It is difficult to express in words what can only be learned by much experience and constant watching; but, in my own experience, there has been generally a period when this condition might be traced with considerable certainty. The pulse is invariably quickened, as soon as the

febrile symptoms are developed, and the heart's
action is necessarily quick too ; but at first
the length of each sound and pause, though
shortened in exact proportion to the rapidity
of the pulse, is not changed in relation to the
other intervals. As soon as inflammation
commences in the heart-structure, the rhythm
which characterizes the succession of sounds
and their relative duration is altered, and the
first sound is no longer distinguished by its
length from the second. The change seems to
depend upon shortening of the systole, rather
than on any lengthening of the diastole, and
is probably due to the well-known influence
of inflammation in interfering with muscular
contraction ; consequently it is much more
evident prior to an attack of pericarditis than
one of endocarditis. If there be any advantage
in the early detection of disease, I think too
much care cannot be exercised in watching for
this sign ; and any unusual rapidity or excitement of the heart's action calls for especial
watchfulness. It is very apt to be forgotten,
in studying physical signs, that the relation
they bear to the symptoms which accompany them must form the basis of our
conclusions, and that neither will alone lead

to a trustworthy diagnosis. It is needless to look for alteration in rhythm as an indication of inflammation, when the heart's action is quiet; when present, in such circumstances, we may feel quite sure that whatever may be its cause, it certainly does not depend upon inflammatory action. In fact, if the interpretation which I have put upon it be correct, it only shows imperfect muscular contraction, but does not show why that imperfection exists.

A case interested me very much some years ago, which will be found in the *post-mortem* records of St. George's Hospital; and occurred at the time when my ear was first becoming familiar with the alteration of rhythm. The patient, a female, was admitted with acute rheumatism of some days standing. The heart's action was laboured and the breathing quick, and the conclusion almost necessarily come to was that the heart was involved. But from her first admission no abnormal sound was ever heard beyond this alteration of rhythm, and in her case it was more than usually marked and distinct. Day after day it was watched, and it was ever present, but the anticipation of further evidence of inflamma-

E

tion was not realised. She sank, as she should
not have done had the heart really been un-
affected; and on *post-mortem* examination, the
pericardium was found universally adherent to
the heart's surface by a layer of recent lymph.
Since that time I have always been in the
habit of regarding it as the effect of altered
muscular movement, and only in this sense
valuable as an evidence of what is going on.
In many cases no doubt it subsides without
further symptoms, and it is then impossible to
say with certainty whether it was caused by
threatened inflammation, which had completely
subsided under treatment without effusion of
lymph, or whether it was dependent on some
other circumstance. But it is a sign to be
watched and studied, as it certainly gives us
more early information of what is about to
occur than any other.

Very shortly after the alteration of rhythm
we can trace by auscultation the presence of
some sound not heard at all in health. It is
well, if possible, to preserve a distinction in
words between the abnormal or accidental
sounds, and those ordinarily heard, by employing such terms as " bruit," " murmur," " friction-sound," &c., so as to avoid any con-

fusion in referring to them. Upon their correct interpretation depends the diagnosis of the form of inflammation which has supervened, whether endo-, or exocardial. In broadly marked examples, the distinction is very easily made by any one whose ear has been properly educated, and I might content myself with referring to books on the subject, without attempting here to enter into any explanation. But there are one or two points which claim something more than a mere passing notice, because, as it appears to me, sufficient prominence is not generally given to them.

In attempting to form a diagnosis in obscure cases, the mind is very apt to be perplexed by the number and variety of indications by which these murmurs are usually said to be characterised, and one or other of them is very apt to catch the attention at the moment, to the neglect of the remainder, when a false conclusion is the almost unavoidable result. It has seemed to me, that the ear ought to be trained to catch the rhythm of the heart's sounds first, and then to group around them any thing abnormal that may be heard, taking note at the same time of any alteration in the natural rhythm, as referred to in the preceding pages.

If this be done, the essential points connected
with the production of each will retain a prominent place in the memory ; and the fact that
they are essential, gives them a definiteness
and a certainty which no other interpretation
of them affords.

Friction-sound is in this respect perfectly
distinct and separate from endocardial murmur.
Their relation to the ordinary first and second
sounds, their time of occurrence and duration,
form the most trustworthy indications. When
these are fully observed and carefully noted,
other signs will almost inevitably guide the
mind of the observer to a true conclusion, but
the first consideration must be the relation
of normal and abnormal sounds to each other
at their point of greatest intensity. The
mechanism by which the various murmurs are
produced differs in detail, but is primarily due
to the muscular contraction of the heart; and
consequently there will be found in each class
of abnormal sounds a relation to the normal
sounds, and yet a distinct character and rhythm
which belongs to themselves alone.

I am not here going to enter into the vexed
question of how the natural sounds are produced;
and there has never been any dispute as to

friction or endocardial murmur. It has been
assumed, without a doubt, that the one is caused
by the rubbing of two rough surfaces together,
and the other by the blood being thrown into
unnatural vibrations in its transit through the
heart. We can therefore readily apprehend
that the rhythm of the two must be essentially different, and must have a very different
relation to the heart's movements. That
this is the case, will be at once admitted by
any one practically familiar with the sounds
of pericarditis.

There are indeed many circumstances which
must be taken into consideration in doubtful
cases. Sometimes indications which are in
themselves the most uncertain, acquire an
unusual importance, but we must be careful
that they do not occupy too prominent a place.
Such are the characters of the abnormal sound
heard, whether it resemble rubbing, grating, or
creaking, in opposition to a blowing or bellows-
sound ; the circumstance of its being heard or
even felt to be very superficial ; the extension
of the cardiac dulness as evidenced by percussion, and absence of respiratory murmur. No
one who has really practised auscultation carefully will pretend to say, from the character of

the murmur heard, what is the cause of its production: neither will much confidence be
placed in the impression that it is produced
somewhere not far from the ear of the auscul-
tator. In the early days of auscultation, such
points received much more attention than in
the present day, and in the writings of French
authors, who followed at no very long interval
after Laennec, the number of strange sound-
ing names invented to convey an idea of the
various *bruits*, tended very much to obscure
the subject of cardiac pathology. Their value
is exceedingly subordinate and secondary, and
many mistakes have, in my own experience,
been made by persons who have endeavoured
to draw conclusions from them. The circumstance that the lungs are pushed aside by
effusions into the pericardium, is a much more
important indication when it is present; but it
cannot exist in the early stages of pericarditis.
I have witnessed such a mistaken diagnosis
when the respiratory murmur was inaudible in
the neighbourhood of the praecordial region, and
the observer was led to conclude that the lung
was pushed aside, while in reality the entrance
of air was only prevented by intercostal rheumatism. An intelligent examination of the

history would in this case have proved what
the absence of very marked dulness certainly
suggested, that there could not have been pericarditis with effusion.

The rhythm of the friction-sound, and the
time of its occurrence, is unquestionably the
most important of the signs by which it can be
distinguished from endocardial murmur. When
the sound is double, as it is in a large majority
of instances, it is remarkable for its uniformity ;
there is no semblance of first and second
sound ; and for aught that friction tells, it is
impossible to say which is systole and which
diastole of the heart. It does not follow the
rhythm of the heart's action ; but, more than
this, there is a consciousness of the two sets of
sounds being wholly independent. If the ear
be first familiarised with the sound of the
heart as heard at a little distance—along the
aorta, for example—and then returns to the
point where the abnormal sound is heard, it is
at once perceived that it neither begins nor
ends with the normal sound, but follows a
rhythm of its own, which does not correspond
in any way except in the frequency of its
recurrence. Even when the friction-sound is
single, as on rare occasions it is found to be,

this independence of the times at which the contraction and dilatation of the heart are traced in the first and second sounds, will still be found a leading feature in the diagnosis of pericarditis.

With endocarditis the indications are exactly the reverse. The rhythm of the murmur is in perfect harmony with that of the heart's action. In the acute stages of the disease it is almost always a single sound; in the earliest always so. And here there is another distinction very marked between the two. It is a matter of absolute certainty, if true pericardial friction be heard, that the disease is present; but it is a very hard matter indeed to say, at any subsequent period, whether pericarditis has previously existed: whereas it often requires the very nicest judgment to decide whether a valvular murmur indicates the presence of endocarditis; but the valve, once seriously implicated, never wholly recovers itself, and maintains a *bruit* at each contraction of the heart to the end of the patient's life; and by this its past existence can generally be readily predicated. This circumstance renders it necessary to consider the whole question of endocardial murmur in connexion with that with which we are more immediately concerned, inflammation attendant on acute rheumatism.

The immediate cause of endocardial murmur is that the blood is thrown into unnatural vibration. In health the structures are so adapted to each other that the blood flows evenly through the heart and vessels without causing any perceptible sound of its own. In disease the relations are changed, either by some alteration in the heart's structure, or by such a modification of the condition of the blood that its passage is marked by its being thrown into vibration in coming in contact with parts of the tissue over which it previously flowed quite evenly. To discriminate these two conditions, is one of the most essential elements of correct diagnosis, and much has been written on the subject, very often leaving it only more obscure than before.

Valvular murmur, the abnormal sound, which is dependent on imperfection of the valve structure, is produced by two processes which are very different from each other. In the one, the blood is thrown into vibration by passing over a surface roughened by disease; in the other, a current is permitted in a reverse direction by the valve being unable to close completely. In the one, the vibration is produced by contact of the blood with the tissue; in the other, by the meeting of two opposite currents of blood. Both circumstances may occur at either the mitral or aortic valve, although practically each is more commonly found at one than at the other. The consideration of the pulmonic and tricuspid valves may be omitted, as they are never inflamed in connection with rheumatic fever, and in their general results they follow nearly the same rules as those of the left side of the heart.

When a murmur is produced at the entrance of the aorta by disease of the valve, in the first instance it accompanies the systole or onward flow of the blood alone, and depends on roughening of the outer edges of the semilunar valves. The deposit does not usually at first interfere with the proper adaptation of the valves during the diastole, and no reverse current is permitted. It is simply a systolic aortic murmur. After a time, probably, the valves become rigid, and do not completely close the aperture, regurgitation takes place, an inverse current meets the blood entering the ventricle from the other side, and a regurgitant aortic murmur takes place. The same result occurs if one of the valves be perforated, still retaining its natural smoothness; but in such a case there is no systolic murmur.

At the mitral valve, on the other hand, the first murmur is always regurgitant. It occurs, like the first aortic murmur, during the systole; it is indirectly caused by the same roughening of the valve, but the roughness only acts by permitting regurgitation. Here an onward murmur is very rare. It requires an extraordinary degree of roughness for its production. This difference seems to me mainly to depend on the force with which the blood is expelled from the ventricle at each contraction of the heart. Probably the shape of the valve is also concerned, but it is manifest that blood driven with the whole force of the ventricle may be thrown into vibration by an amount of obstruction which would be insufficient for its production when propelled only by the degree of force developed in the contraction of the auricle.

So also, a very slight amount of roughness on the edges of the mitral valve renders it incapable of withstanding the pressure of the blood during the ventricular contraction, and regurgitation gives rise to a distinct sound; whereas the same amount of roughness on the aortic valves does not incapacitate them from preventing the return of the blood when only driven against them by the resiliency of the aorta itself.

If regurgitation be proved to exist, there can be no question that the valve is imperfect. But if a murmur accompany the onward flow of the blood only, it may depend on change affecting the fluid itself without derangement of the valvular apparatus; such murmurs have consequently been called functional. Let us now see if there be any positive rules for diagnosis in each case.

The systolic aortic murmur is that which is most likely to be simulated by a functional one. We may at once set aside any distinction that may have been drawn from the character of the sound itself as quite fallacious in doubtful cases. But it has been alleged that a functional murmur may be traced most distinctly upwards towards the left shoulder, while a valvular murmur is heard more clearly across the sternum. This statement I believe to be very apt to mislead. It is partly true

and partly false—true that a murmur heard
over the second intercostal space on the left
side is generally functional; false that a
murmur heard across the sternum at the
second right intercostal space is generally
organic.

The facts are very simple, the theories based
on them often false. In the one position the
movement of the blood in the pulmonary
artery is most distinctly recognised, in the
other the movement of the blood in the aorta;
and in either case there must be some unusual
vibration to make it audible. Disease of the
pulmonic orifice being extremely rare, a murmur in the pulmonary artery is almost always
functional in its origin. Disease of the aortic
orifice, being not uncommon, an aortic murmur
may be either functional or organic. My
own experience leads me to say very positively
that when a murmur can be traced some distance along the aorta there is always present
a changed condition of the blood. The vibration is by this means propagated along the
artery equally when it takes its rise in some
obstruction at the aortic orifice, and when it is
due to the passage of unhealthy blood from
a large cavity into a narrow tube, as is the

case in simple functional murmurs. I am quite certain that the very roughest and loudest murmurs are sometimes scarcely audible on the right side of the sternum, and that the most conclusively functional murmurs are often louder there than at the base of the heart.

Whence then are we to derive our diagnostic signs ? First, we have a very valuable one in the power of producing murmur by pressure with the stethoscope in cases of functional derangement, both in the carotid artery and in the veins of the neck. Secondly, an organic murmur being produced by a distinctly local cause, which can easily be covered by the end of the stethoscope, has its point of greatest intensity clearly marked just after the blood has passed the seat of disease. With the help of these two indications an educated ear, guided by a cautious consideration of all the circumstances of the case, will seldom fail in determining whether functional disorder or organic change is the only cause of the murmur, or whether the two are co-operating in its production. Nay, more, it is sometimes possible to say with tolerable certainty that the sound which is heard over the right intercostal space is an indication neither of disease of valve nor of blood-change, but of disease of the coats of the vessel. The result of very extensive auscultation of the heart has been to force on my mind the conviction that functional murmurs are especially vague in their site, and that it is almost impossible to fix on any spot where they attain a maximum of intensity ; whereas organic murmurs always have a point of origin and of greatest intensity, with the single exception, perhaps, of some regurgitant aortic murmurs.

Much confusion seems to me to exist in the minds of writers on this subject, because of their assuming that a blood-murmur must be produced at a valvular orifice. If the valve be sound it has really nothing to do with it, and any theories which attempt to explain its production by alterations in the form of the heart, or by perverted innervation or disordered contractility of the valve are really useless. It matters not whether the patient be blanched in appearance or have a florid aspect ; if a blood-murmur can be produced by slight pressure on the carotid, a similar murmur must be produced in the heart. There is no roughness or unevenness of surface in the carotid, there is merely a narrowing of the passage, so there need be no roughness or unevenness at the outlet of the heart, the mere narrowing of the channel must of itself be sufficient. And when the tendency to be thrown into vibration is great, it need not be restricted to that situation. In point of fact, the vibration is very generally produced before the blood reaches it. Careful auscultation will generally trace it in the cavity of the ventricle itself, probably excited there by the contact of the muscular bands which everywhere project from the surface.

Attention has been of late turned to this point, and various suggestions have been made as to the mechanism of murmur at the apex which was not dependent on organic disease of the mitral valve. Many years ago it came under my own consideration, before I had seen any notice of the subject by others. I had been taught that a murmur at the apex indicated mitral regurgitation, and that mitral regurgitation implied valvular lesion, and I was forced to unlearn this teaching by unbending facts. In its place the following conclusions were arrived at from the observation of a large number of cases, and they are here suggested for the consideration of others, though they cannot be regarded as proved. First, that mitral regurgitation does always imply insufficiency of the valve, but not always disease. Secondly, that a murmur at the apex does not always imply mitral regurgitation.

The insufficiency of the valve in some cases of hypertrophy depends on the want of adaptation of the valve to the altered condition of the heart, while it seems, in others, to be caused by shortening of the *chordæ tendineæ*. It is therefore not impossible, though to my mind highly improbable, that in chorea irregular action of muscle may cause imperfect closure of the valve, and permit regurgitation.

But if there be not regurgitation, how are we to account for murmur at the apex? The answer must necessarily be hypothetical. In examining with great minuteness all the blood-murmurs that came before me, the first and almost inevitable conclusion was that they did not present the character of having any definite point of greater intensity. Further, that though diffused, there was very generally a limited area in which, though not quite local, they were heard louder than elsewhere : and that this was sometimes at the second intercostal space, sometimes over the centre of the heart, sometimes towards the apex. It was then

F

noted that these regions of greatest distinctness corresponded to intercostal spaces, and I
arrived at the conclusion that this was one of
the localising circumstances. It was further
observed that when the murmur was heard
very distinctly at the third, or more especially
at the fourth intercostal space, and less distinctly at the second, the breathing sound was
audible at the higher point, and absent over
the surface of the heart; and that the
further down the breathing was audible, the
more did the blood-murmur seem to be produced at the apex. In short, the conclusion
arrived at was that though functional murmur
is most commonly heard when the blood
passes from a cavity to a tube, yet it may also
be produced in the cavity itself, and that the
chief localising circumstances are these: first,
its tendency to be produced at the commencement of a tube, whether aorta or pulmonary
artery; second, its being loudest when the
lung did not interpose between the blood and
the ear of the listener; thirdly, its being more
readily transmitted through the intercostal space
than through the rib.

The murmur produced by mitral regurgitation before dilatation and hypertrophy of the

heart have set in, is so remarkably local and definite that I look back with surprise to the time when a functional murmur loudest at the apex was mistaken for it. It is only in the early period when the size of the heart is still normal that there can be any question : and at that period, if careful examination be made, the ear cannot fail to come on a small defined space, generally not more than an inch from the apex-beat where the bellows-murmur, the old *bruit de soufflet* of the early French auscultators, comes out with a sharpness and distinctness which is often lost after other changes have taken place in the heart. When hypertrophy and dilatation have supervened, especially if accompanied by irregular action of the heart, the mitral murmur may become very indistinct.

I have been led to discuss this matter more fully, because many years ago the possibility of functional murmur occurring as the effect of acute rheumatism was first brought distinctly before me ; and I have several times since had to advise with medical men to whom the possibility was quite unknown, and who looked upon a cardiac murmur of any sort as of grave import in such circumstances. The case to which I refer occurred in St. George's Hospital. A healthy young woman was brought there, suffering under severe rheumatic fever. She had only been ill one or two days before admission. The pulse was quick, and the heart's action correspondingly excited, but there was no alteration of rhythm nor morbid sound of any sort to be heard on her admission. A few days later a murmur was observed at the base of the heart, which was persistent, and increased in intensity for two days more. It was, without further question of its origin, set down as the result of rheumatic endocarditis, though no doubt, with more careful analysis, it would have been found not to have the true character of a valvular murmur. On the third day of its existence she was seized with delirium, and sank rapidly. The heart was opened after death, and not a trace of fibrinous deposit was found on the aortic valves. The conclusion was inevitable that the alteration of the blood during the existence of rheumatic fever may be of such a kind as to give rise to blood-murmur. In no case which has come under my observation has it been so distinctly capable of proof as in that just related. Several instances have subsequently presented them-

selves in which a murmur actually present at
the time of examination was set down as
functional, and the result has justified the
prognosis ; but it has not always been so certain that it arose during the period of the attack.
One case may be scarcely considered sufficient to
prove such an assertion, were it not that the
knowledge we possess of blood-change occurring in acute rheumatism renders the result
not improbable.

Instances of this kind are no doubt exceptional; and in a majority of cases cardiac
murmur is due either to existing endocarditis,
or to previous disease of the valves. The most
open to question are those occurring at the
base of the heart, those met with at the apex
ought never really to perplex the listener. A
very small amount of deposit on the mitral
valve is sufficient to prevent the perfect adaptation of its two flaps ; and when we reflect on
the force with which the ventricle contracts, it
does not seem surprising that any slight imperfection should permit the blood to escape
through the chink formed by the valve. When
such a defect in the valvular apparatus exists,
I may repeat that there is always a circum-
scribed spot where the sound produced by the

regurgitation of the blood into the already
filled auricle can be most clearly and unmis-
takeably traced. This fact cannot be too
steadily kept in view. No amount of roughness of the first sound—no alteration of its
character, even when the sound at the base
differs from that heard at the apex, ought ever
to be taken as evidence of mitral regurgitation
until the short sharp whiff be distinctly made
out.

The limited range over which this true
mitral murmur can be heard is very remarkable, and leads not unfrequently to its being
overlooked. Its limit and position are liable
to occasional variations, which sometimes increase the difficulty of its recognition. My
efforts to ascertain the causes of such variations
have not been successful ; I will therefore only
add that the most ordinary position is nearly
on a level with the apex-beat, and about one inch
nearer the sternum. When such a murmur is
clearly made out there can be no question of
mitral regurgitation. All other modifications
of the first sound towards the apex may be due
to some other cause ; and were I to judge by
my own experience, I should say must be so if
the pure regurgitant murmur just described

does not exist. This explains the circumstance so often alluded to by authors, that a murmur heard towards the apex may turn out to be pericardial friction-sound in place of endocardial murmur. It has evidently never been in such cases the true sound of mitral regurgitation, and the observer has been only misled by the sound being a single one.

One case, however, I must record, which proves, to my thinking, that mitral regurgitation does not necessarily imply disease of the mitral valve. The patient was a young person who had very severe pericarditis during an attack of rheumatic fever—such, indeed, that the conclusion was very fairly come to that the pericardium had become wholly adherent. Some days after the last trace of friction-sound had disappeared, a very distinct mitral murmur was heard. Of its character and meaning there could not be two opinions, and it was persistent. It was therefore assumed that the poor girl had unquestionably suffered from endocarditis as well as pericarditis. She subsequently left London for change of air, with the mitral murmur unaltered, and returned to town after two or three months' absence. No treatment was adopted during the interval, but when she

returned the heart was perfectly free from murmur of any kind.

It is extremely difficult to propound any
hypothesis which may serve for the explanation of such a fact as this. The most reasonable, perhaps, is that the adhesion of the
pericardium had such an effect on the muscular
tissue that the *chordæ tendineæ* served to keep
the valve open for a while, as occurs sometimes
in enlarged hearts ; and that gradually the
organ became accustomed to the new condition, and the valve, which had never been
diseased, returned to its normal position. It
goes for very little in such a case, that during
the early part of her illness the valvular murmur
was not heard, because the friction-sound was
so loud ; the presumption, however, is strong
that it would have been heard in the last few
days of the friction if endocarditis were
present, even though in the earlier stage it
might have been inaudible. I confess that my
own experience does not justify me in coming
to the only other conclusion which might be
supposed to offer a solution of the difficulty—
viz., that there had been complete recovery
from inflammation affecting the mitral valve.
It is a sad result to arrive at that such cases

never wholly recover, but at present our efforts must be directed to ward off the attack if possible, and to mitigate the subsequent symptoms when it supervenes.

What, then, is the future of such cases? The earlier observers were in the habit of forming a most gloomy prognosis. A few years were supposed to be the extreme limit to which they were likely to extend. At one time I saw a patient occasionally who had been condemned by the late Dr. Hope nearly twenty years before, on account of a mitral murmur which was left after an attack of rheumatic fever. His experience only taught him that very soon after valvular disease was fairly established, further changes were usually developed in the heart's structure, and all the train of symptoms which disease of the heart is supposed to imply followed in due course. But this girl had taken his warning in good part, had fortunately been able to lead a quiet sedentary life; and during two or three years, when I was asked occasionally to see her, I did not find any evidence of increased disease. Such care will generally retard the progress which disease makes when left to itself, but in few instances is the result likely to be so gratifying. In the case of the poor, who are obliged to work for their maintenance, who are exposed to all weathers, and to whom care for their health is almost an impossibility, it usually advances with rapid strides.

There are many other circumstances to be taken into consideration before a fair estimate can be made of the probable duration of life with a damaged heart, and these will more properly come under consideration at a future time. In some instances the condition may undergo some improvement, in others it remains nearly stationary, while in a large number it tends to become continually worse. In the natural process of recovery from pericarditis, either total or partial adhesion takes place, or a small amount of independent deposit remains on one or both surfaces of the sac which contracts no adhesion whatever. In the latter event, the roughened surface gradually becomes smoothed down, great part of the lymph is absorbed, leaving only a thickened membrane behind which does not seem to act at all prejudicially on the organ. If adhesions are found, the damage is very much in proportion to their extent, so long as the two surfaces are not wholly united. No condition produces enlargement of the heart with greater certainty than extensive adhesion combined with valvular lesion.

In cases of endocarditis there is a similar process of repair gradually carried on for some time after the attack has passed away. The lymph beads on the edge of the valve are removed by increased vascular action in the minute capillaries; but along with this a distinct thickening and hardening of the valve structure takes place, passing probably through the successive stages of tumefaction, exudation, and shrinking. At first there is, perhaps, an amelioration of the symptoms of disease, and then a gradual increase may be observed. At the aortic valves, for example, the obstructive disease is first slightly diminished, and then by the shrinking of the valve regurgitation is permitted, and one of the most certain causes of enlargement is established. The recoil with which the blood flows back into the ventricle meeting the onward current from the auricle, makes a constantly increasing demand on the propelling power of the organ, while at the same time it is always distended with a large amount of blood, and so hypertrophy and dilatation occur together. Mitral disease in its worst forms, leads to similar over-action. When the valve remains so widely open that an amount of blood is thrown back into the auricle at each contraction nearly equal to that which is propelled through the aorta, the requirements of the system, especially during prolonged muscular exertion, compel the heart to act energetically and rapidly, and as a consequence, enlargement must of necessity follow.

It is when changes of this kind have taken place that cardiac disease really presents its formidable character. A mere *bruit* while the heart is quiescent in its action, and unaltered in form is comparatively a small matter. It points to the future, it tells what may happen, what almost certainly will happen; but as yet danger is distant, and escape perhaps possible. But when the organ is enlarged, the aspect of the case is wholly changed, and it is difficult even to retard its progress.

This is only one instance out of many in which the remedial agency of nature works in the end its own destruction. Disease demands an increased propelling power, and the more the muscle acts to produce this effect, the more is its power increased, just as the arm of the blacksmith becomes proportioned to the labour which he has to sustain. But then this increase cannot be kept within due bounds, and in proportion as the action goes on so does greater increase become necessary. Local congestions of various organs assume a constant character, in consequence of the abnormal impulse of the blood, and these in their turn help to make the circulation still more laboured. At length impaired health renders the blood more watery, congestion of the kidney perhaps interferes with the secretion of urine ; serous exudation takes place in the areolar tissue, and a series of symptoms is begun which may end only with life.

Sometimes a more sudden termination awaits the sufferer. If his blood be not impoverished by lingering illness it may in a moment of excitement be driven with more force than usual to the brain, and an apoplectic seizure follows. In other cases, mental emotion or sudden muscular strain produces violent irregular action of the heart, and the circulation suddenly interrupted never recovers its normal regularity. Sometimes the heart's action ceases instantaneously ; sometimes life is prolonged for a considerable period, all the while attended with much suffering.

In all these details the origin of the disease
in rheumatic fever does not impart any distinctive character to the cardiac affection. If
the valve be insufficient it matters very little
whether, in the first instance, it commenced as
inflammation, with deposit of fibrine on its edge,
or whether degeneration of its tissue had taken
place with atheromatous or calcareous deposit.
Two circumstances only are to be remembered,
and these bear rather on the probable progress
of a case seen in the earlier stages, than of one
in which muscular change has already taken
place.   First, it is in the earlier period of life
that rheumatic affections of the heart begin,
when the circulation is more active, and the
habits of life are such as are more likely to call
for increased muscular action, if there be any
obstruction to the circulation.   Secondly, there
seems no doubt that a heart which has once
suffered from endocarditis is more liable to
deposit of lymph at subsequent periods than
one that is healthy.   Degeneration once begun
very surely goes on to further stages of development, and fibrinous deposit with equal
certainty leads to contraction of the valve tissue.
The atheroma may be accompanied by thickening and comparative rigidity, and the whole

valve may be ultimately converted into a calcareous mass, which will thus render it perfectly useless; but such a result is slow of
attainment, and it is long before insufficiency
is a marked feature of the disease. Whereas
the contraction of the edge which invariably
follows inflammation with much deposit of
fibrine, very soon gives rise to imperfect
adaptation of the valve, and the return
of a large amount of blood through its half-
closed orifice. In both these respects the
anticipation of more rapid progress in rheumatic cases is fully justified by the results, as
the duration of life is much shorter in rheumatic
affection than any other.

CHAPTER IV.

———◆◆◆———

Inefficacy of treatment to stop the progress of Cardiac Disease —
Prevention by Alkaline treatment — No arrest of the Rheumatic affection — Recovery without treatment — Injurious
effect of Opium — The limited operation of Mercury —
Liability to relapse — The "supporting" plan of treatment
— Its fallacies — Evils resulting from Acidity — Treatment
of Inflammation — Palliatives.

THE means at our disposal for the relief of
disease of the heart are mainly palliative. A
knowledge of the actual condition of the organ
is essential to determining what measures are
most likely to be beneficial; but when all has
been done that art can do in the matter, the
integrity of the valve spoiled by disease cannot
be restored, nor can we reduce the size of the
hypertrophied heart. The chief object of treatment must be to prevent these conditions from
becoming more injurious to life, but much may
be done in the early stages. With the knowledge of the tendency of acute rheumatism to

attack the heart, has gradually grown up the more important knowledge of the means best calculated to prevent such an occurrence, and so avoid the damage which must of necessity follow from endo- or pericarditis. Formerly men were satisfied to view these changes as almost unavoidable, to watch their advent, and, when discovered, to ply the various means which art had seemed to point out as best calculated to remove them after their incursion. It must, I fear, be confessed that the whole train of antiphlogistic remedies — bleeding, blistering, leeches, calomel and opium, and antimony have been miserable failures. The inflammation once begun may have been in some degree restrained or subdued, but the damage was already done before they could be brought into operation, and a life of misery entailed on the patient which such measures were powerless to avert.

More recently the attention of physicians has been turned with the happiest result to counteracting the one prominent symptom in rheumatic fever, viz., the extraordinary degree of acidity of all the secretions, which unmistakably points to an acid-generating power as one of the most definite phenomena of the

G

disease. In seeking to remedy this condition, the unlooked for result has been obtained of preventing to a great extent the liability to cardiac complication. It is a matter now placed entirely beyond question that when the secretions are maintained in a state of alkalescence by the free administration of potash, the tendency of the disorder to excite inflammation of the heart is almost entirely obviated.

This result is one which could not have been arrived at *à priori*, and we are still unable to account for it, though we must admit that the administration of alkalies is a most rational method of dealing with acute rheumatism. We are not yet in a position to say that it exercises any influence in curtailing the duration of the disease. It certainly does not at once arrest its progress, or prevent other joints from becoming in succession affected by the disorder. It does not even appear to bring very great relief to the sufferer; but in spite of all this it does most incontestably prevent inflammation of the heart. Of course it is by no means an easy matter to obtain statistics which will prove whether the disease is less painful and less protracted under the alkaline treatment; but it is by no means difficult to show that the proportion of heart ailments is very much reduced by its employment.

We know that by pouring into the blood a large excess of alkali which passes off by the skin, the kidneys, and the bowels, we can deprive these secretions of their acidity; and if acid be present in the blood, we know that the alkali must combine with it there, and neutralise it. In doing this we are fulfilling one of the most natural indications which medicine can ever be called on to fulfil, and when the treatment is fully carried out the proclivity to heart disease is almost entirely obviated. The excess of alkali clearly cannot be injurious so far as the disease is concerned, and it is manifestly beneficial in its protecting influence over the heart.

Rheumatic fever is of itself by no means a formidable disorder. It is attended by much pain, perhaps by more pain than any other with which we are acquainted; but apart from the affection of the heart its natural tendency is to convalescence sooner or later. It may indeed prove fatal by inducing a condition of brain which for want of any better defined character has been supposed to be produced by metastasis. The cases are so few that it is impossible to speak dogmatically concerning them, but the alkaline treatment may also be regarded with confidence as the best means of preventing such a termination of the attack.

Very recently an experiment has been tried at Guy's Hospital, which proves in a very remarkable manner the rapid convalescence of patients to whom no drugs are administered, and the great liability to disease of the heart in such circumstances.* Of forty-one patients there were only eleven in whom the heart is known to have been perfectly healthy; but in two cases, no record having been kept, we are fairly entitled to assume that in them also it was free from disease. The teaching of this experiment is very full of meaning. There can be no longer any doubt that rheumatic fever left to itself will pass through a regular and definite course, which usually terminates either in complete convalescence or in recovery with some amount of permanent damage to the heart. According to the statement just quoted, only from twenty-seven to thirty-two per cent. of the cases wholly escaped. It has been already shown that a *bruit* may often be heard which is caused simply by altered blood, and therefore we

* 'Guy's Hospital Reports,' vol. xi. p. 415.

should not be justified in assuming that the proportion of cardiac affections was sixty-eight per cent., although the reporter, evidently not anxious to make out an unusually favourable result from Dr. Gull's mint-water treatment, has not thought it necessary to separate the doubtful cases. The highest recorded percentage is that of Bouilland, in Paris, when he assumes that eighty-six per cent. of his cases had organic disease of the heart.[*] This no doubt includes a large number of anæmic murmurs; and when it is remembered to what an excess he carried the practice of venesection, it may probably be assumed that the proportion of blood-murmurs was much greater among the cases recorded by the French physician than among those occurring within the last two or three years at Guy's Hospital. But whatever allowance may be made for *bruits* which do not indicate serious mischief, the proportion of heart affections is in either case very large. In the records which I published some years ago I ventured to separate doubtful cases from those in which there was no doubt

---

[*] Bouilland ' Traité Clinique du Rhumatisme Articulaire,' p. 143.

[†] ' Med. Chir. Trans.,' vol. xxxv. p. 18.

in my own mind of the existence of disease.
The result was that of one hundred and fifty-
two cases of acute rheumatism, sixty-four had
no cardiac affection ; sixty-seven gave evidence
of recent or old standing disease ; while twenty-
one were recorded as doubtful. Adding the
two latter together we obtain for comparison a
percentage of forty-two free from disease, against
fifty-eight in which either friction-sound or
bruit was heard. This result was obtained
from cases treated on the old methods, which
are thus shown to have some advantages over
the do-nothing system of Dr. Gull, although
falling very far short of what can be done by
alkaline treatment.

The result so far is just what previous
experience would have led us to anticipate.
The shrewd observers of a day gone by
had already arrived at the conclusion that
treatment had very little influence over the
duration of rheumatic fever, and that it was
impossible by any means they had tried to cut
it short. The once celebrated saying that
" six weeks " was the best treatment for the
disease, proves indeed to have gone beyond the
truth, and the effects of mint julep will bear
comparison on the point of duration with other

methods. The all-important question, however, remains, one which, more than any other,
influences for good or ill the whole after-life
of the patient, viz., What is the result of a
comparison between the alkaline and the ex-
pectant treatment as regards the heart? The
only recently published statistics to which I
can turn are those of Dr. Dickinson.* They
are very short but instructive, and on one
point only can they be compared with those
of Dr. Gull. Out of forty-eight patients subjected to the full alkaline treatment, only one
had pericarditis, commencing after treatment
was begun. In the forty-one cases at Guy's
Hospital there were six cases of pericarditis.
It is true that in several of these the inflammation had begun out of doors, but as the object
of the report is to give an idea of the natural
history and progress of the disease, they cannot
be excluded in this enumeration, though they
ought to be, and must be, when giving the
results of any particular treatment which has
for its object the guarding the heart from an
attack of inflammation. Before admission few
patients are subjected to full alkaline treatment, and many are consequently admitted with

* ' Med. Chir. Trans.,' vol. xlv. p. 350.

recent inflammation, which cannot be arrested by remedies.

The real value of the saturation of the system with alkalies can only be thoroughly tested by an exact analysis of every case, to ascertain whether there was any indication of inflammatory action before the treatment commenced. The result of such an inquiry may, I hope, some day be laid before the profession; in the meantime, I may say with confidence, that in my own practice the lighting up of inflammation of the heart, after the system has been brought fully under the influence of the alkali, is one of the rarest events in pathology. This experience, to one who was in the habit of watching the condition of the heart in all the cases of rheumatic fever in the same hospital fifteen or twenty years ago, is most striking, and needs no confirmation from statistical calculations.

There is however an element of fallacy in all such investigations, which applies with special force to acute rheumatism, viz., that there is no standard by which we can measure the intensity of the disease. The importance of such a consideration must be apparent from what has been already stated of the relation between the intensity of the disorder and the liability to cardiac affection. One observer will naturally include a larger number of rheumatic patients in the list of acute cases, another will limit the term more rigidly, and class more of his cases as sub-acute. It is not easy to avoid this source of error in making a comparison of different modes of treatment. In addition to this, the mean age of the patients should be stated, as there can be no real contrast between a series of cases occurring under thirty, and one in which the patients are all over forty years of age. In the one case the liability to heart affection is great, in the other it is almost inappreciable.

Much harm has probably been done by the over-anxiety of practitioners to give relief to the sensations of the patient. What more natural, when such pain was to be endured, than to endeavour to lull the sensibility of the nervous system by opium? And yet, perhaps, no drug is more likely to do harm if administered too freely. The secretions are all more or less liable to be checked by its influence. The bowels especially are locked up, and the urine, at all times scanty, becomes

still less abundant, and more loaded with
effete matter, which the system is unable to
throw off. Then, too, the increase of suffering,
produced by every movement, predisposes both
patient and attendant to abstain from the
use of purgatives, and the blood necessarily
becomes still more vitiated, more charged with
waste and noxious constituents. It is, therefore, not surprising to find that cardiac
complications are more frequent when opiates
are freely employed than under any other
circumstances. Not that the use of opium
can be wholly dispensed with, but it must
be used with judgment, and care must especially be taken to maintain full action in all
secernent organs, and to bring the character
of the secretions as closely as possible up to
the healthy standard.

Mercury, so often misused, so falsely
praised for results to which it did not contribute, so frequently blamed for evils which
it never had the power to produce—is often
of essential service in the treatment of rheumatic fever. Not, indeed, administered blind-
fold, as was at one time the custom, in
constantly repeated doses, till the system became poisoned by its presence, and salivation

attested that it had been already given in
excess ; but employed with the rational view
of correcting and improving the secretions of
the alimentary canal. A very limited experience is sufficient to convince any one how
much good may be done by mercurials when
prescribed in suitable doses. So unmistake-
able is the evidence of their value, that one
may be tempted to trust to them too much, to
use them too freely. The excess is probably
the worst evil of the two ; but there is rather
a tendency in the present day to run into the
opposite extreme, and to miss the chance of
good which may be fairly looked for from their
employment. Mercury does not hold the
same position with reference to rheumatism
that must be assigned to alkaline remedies.
It does not serve to guard the heart from
inflammation, and although so constantly used
to combat the inflammation after it has
begun, it may be doubted whether it exercises any great influence in this direction.
It cannot be shown that patients under the
influence of mercury recover more quickly than
under any other treatment; and, as the
experiment at Guy's proves, recovery alone
cannot be taken as any test of its value. Those

who have recovered under its administration
would, as it appears, with equal certainty have
got well without a grain of the metal having
been given, or any other treatment having been
adopted.   As an aid to other remedies it has
great claims on our attention.   I have certainly seen the tongue become rapidly clean,
and the secretions once more assume their
natural character in a way which I believe
could not have happened had it been withheld.

   When the acute stage has passed, many
other remedies fulfil very important indications in enabling the system more thoroughly
to recover from the attack ; but with these we
are not now concerned, as the liability to inflammation of the heart is then at an end.
Throughout the period of danger, the sheet-
anchor of the physician is the possibility of
neutralising the acid generated in the system.
If the treatment be left off too early there is
a great probability of the recurrence of acute
symptoms.   My own experience of the liability
to relapse in such circumstances, confirms a
conviction which is perhaps incapable of more
direct proof, that, invaluable as the alkalies are
in guarding the heart from an attack of inflammation, the advantages of their employ-

ment are not thus limited: but that rheumatic fever is really less prolonged and less painful in the acute stage, and less liable to degenerate into a lingering disorder, when alkalescence of the secretions is maintained after the period of danger to the heart has passed.

No greater mistake in regard to treatment could have possibly been made, as it appears to me, than the application of the "supporting plan" to acute rheumatism. Theoretically false, and practically injurious, the authority of a great name has led to its adoption in a more or less modified form by many who have not perhaps sufficiently considered the subject, but have followed a fashion which, passing from the one extreme of excessive bleeding, has rushed blindly into the other of excessive stimulation. The need for support under illness is just in proportion to the danger of the patient sinking from exhaustion during its continuance. Such a state is not the ordinary course of acute rheumatism. It has been shown that recovery is the natural termination of the disorder ; and were it not for its tendency to attack the heart, it might, with perfect safety, be left to run its own course— perhaps a more painful and lingering one, but not one at all dangerous to life. Whence, then, the need of supporting the patient by copious supplies of food and stimulants?

There is, no doubt, a period in its history when the disease in many constitutions tends to assume a chronic character; and if the tone of the system be very much lowered such a tendency is more marked. That it is the duty of the physician to guard against this source of danger to the patient no one will deny; but that it can be averted by the free use of wine and brandy is a conclusion opposed to all our knowledge of the nature of the affection, and is constantly contradicted by experience at the bedside. Patients will often derive benefit from other forms of nutriment who cannot bear the smallest amount of stimulants, and even actual starvation is better than excessive supply.

In the early stage of the acute attack there is often a full and bounding pulse, such as in earlier times led to the free use of the lancet; and if we now think it unnecessary to bleed, this well-marked feature points out at once that starvation cannot possibly do harm, and is really likely to be beneficial. I cannot conceive that, in direct opposition to such an indication, there

can be any justification for the administration
of stimulants. My own experience most decidedly confirms this rational inference. When
stimulants have been administered early, the
disorder has been generally lingering and
obstinate, and the enforcing of a few days'
starvation, especially in persons of a plethoric
habit, has been attended with the happiest results. It is very difficult in the practice of
medicine to point to consequences, and refer
them with any degree of certainty to the actual
causes from which they have sprung. But
when a rational conviction has been fairly
arrived at, and the anticipation has, in a
majority of instances, been corroborated by
the facts as afterwards observed, there is, at all
events, some ground for the belief that the
proposition was correct. Could we assume
of rheumatic fever as of typhus, that it was
essentially of an adynamic type, and that the
chances of recovery were only in proportion to
the powers of resistance, it would be reasonable
to allege that the patient must be adequately
supported, and the only question would be as
to how that support should be administered.
But if, on the contrary, the disease does not
endanger life, and the chief risk consists in the

chance of inflammation of the heart occurring while the febrile action is high, and the patient has not yet been lowered by its persistence, it must be wrong to add fuel to the flame. Even when the time comes for administering nourishment, and relieving the exhaustion caused by continuous febrile action, I am sure that great caution is necessary, and that it is of much importance to adapt its character as well as its amount to the actual wants of the patient, so as not to re-excite the acid-forming tendencies which are so prominent throughout the attack.

However distinct gout and rheumatism may be in their essential elements, it is a most remarkable fact that in both there is a corresponding tendency to the generation of acid. The acid of gout is not that of rheumatism, but the substances which taken into the system promote acidity in the one disorder are those most likely to do so in the other. Necessarily the acid is in either case formed at the expense of some other product. That which should have gone to the formation of new tissue, or to the maintenance of secernent action—the two processes which make up the sum of organic life—is converted into an abnormal product which must be eliminated. To meet this demand the tissues are wasted, and they must in their turn be replaced by food; but on its due selection depends very much the arrest or continuance of the abnormal acidity. For example, a patient in the first stage of convalescence complains of excessive weakness, and longs for permission to indulge in his ordinary habit of taking stimulants along with the simple food which has been prescribed. How often does the yielding to such a desire entail a recurrence of the more acute symptoms, and retard instead of promoting ultimate recovery!

There is no doubt a "poor" rheumatism as well as a "poor" gout, a condition in which the system already lowered by existing disease, will not bear the additional depression of actual starvation. This form of rheumatism does not lead to inflammation of the heart, and therefore scarcely falls under our consideration at present. It may be remarked, however, that such patients must be properly sustained, and that convalescence may be indefinitely postponed by the want of suitable nourishment. In this, as in the more acute forms, the greatest care is needed to avoid supplying those ma-

terials which the morbid condition of the
patient will convert into acid, with injurious
effect on the progress of the case. On the one
hand the very simplest food may produce
acidity when the digestion is impaired, and
may require the addition of a stimulant to
prevent such a result; and on the other, the
stimulant if not well borne may set agoing
that irregular process of assimilation by which
acid gets into the blood and re-excites the
symptoms of rheumatism. The judicious physician will in such circumstances endeavour to
rectify the stomach ailment while avoiding the
tendency to recurrence of the rheumatic affection, and may often have recourse to some
stomachic remedy in preference to an alcoholic
stimulant. It is remarkable how many of the
popular remedies for gout or rheumatism have
combined some agent of this class in their
constitution; and it is certain that without
attention to the state of the digestive organs
it is quite impossible to prescribe remedies
suitable to either disorder.

The employment of the means already indicated will, without doubt, in a great majority
of instances save the heart from ultimate
damage. But there are many cases in which

the disease has been either injudiciously managed or left to itself, and inflammation has already commenced when the patient is first seen; perhaps valvular lesion is then only discovered which has taken its rise in some former illness. At no very distant period in the history of medicine, a mere suspicion of inflammation was sufficient to call into requisition all the most powerful agents in the treatment of disease, furnished by experience; and no sooner was it known that disease of the heart commenced as a local inflammation during an attack of acute rheumatism, than all those means were immediately brought to bear upon its earliest stages. Sad experience has taught us how powerless such agencies are to put a stop to the action when it has really been excited. Copious general and local abstractions of blood with absolute starvation, so vaunted by the French school of a former day, have utterly failed, though they were supposed to rest on a sound pathology and a correct appreciation of the relations of hyperinosis of the blood to inflammatory action. Calomel and opium in large and frequent doses, and tartar-emetic, each of them very powerful agents, have proved alike inefficacious in arresting its progress; and it would seem that the interference of art must be restricted to the employment of measures calculated to carry the patient safely through the inflammatory attack, and perhaps to modify to some extent its present severity and ultimate results. No special rules for guidance in such circumstances can be laid down, as almost every case presents peculiarities of its own, which can only be noted by actual observation. The alkaline treatment must be persevered in, and counter-irritants applied over the region of the heart. Not unfrequently the tongue becomes cleaner under the use of small doses of mercury two or three times a day, when no attempt is made to obtain what are called its constitutional effects, and an amelioration of symptoms generally follows. While therefore having no confidence in calomel and opium as a specific means of arresting cardiac inflammation, I may say that there are few cases in which mercury is not needed for correcting vitiated secretions and restoring healthy action.

Experience does not justify the expectation that the solvent powers attributed to certain substances, such as iodine and mercury, afford much aid in removing the fibrin which has been deposited upon the inflamed valve, and which impairs its efficiency. I have given them singly and in combination for long periods, after the acute stage had passed, at a time when the natural process of repair is most likely to take place, but I have not observed that the intensity of a murmur has been even diminished by such treatment. Sooner or later the patient invariably lapses into a condition of chronic disease, a subject to which we must again recur.

CHAPTER V.

THE RELATION OF GOUT TO CARDIAC DISEASE.

The difference between Gout and Rheumatism in this respect — The acid formation — The age of the patient — Rheumatism not hereditary — Liability to recurrence — Independent causes of heart complication — Association through Bright's disease — Characters of Gout when present in such cases — Priority of influence — Mutual origin in Blood-change.

It is very well known that an attack of acute gout is not liable to be complicated by inflammation of the heart or pericardium, and the lesions of the organ which are found in cases of chronic gout, do not generally exhibit an inflammatory origin. Why this is so, it is perhaps not very easy with our present knowledge to explain. When the two diseases are contrasted with each other, there is found to be the same prominence of local symptoms, along with the presence of an unquestionably general disorder : in each the local manifestation consists in pain, accompanied by heat of the joints, redness and swelling; and the general disorder is chiefly characterised by a tendency to acid formation. But the acid generated in the two affections is not the same. On this point medical authorities have been long unanimous, though the fact has only of late years been clearly brought out. The urine of the patient labouring under acute rheumatism may present a very copious sediment of reddish lithates, but it is not uric acid which is found in excess in the system. The amount of the deposit is regulated by three circumstances: first, there is a general febrile disturbance which gives rise to excessive molecular changes, and a more copious excretion of effete matter; secondly, the relative quantity of water in the urine is very much reduced as a consequence of inflammatory action and copious diaphoresis; thirdly, any form of acid in the secretion causes the uric acid salts to assume a more insoluble form than when the urine is neutral or an excess of alkali is present. On the other hand, a patient suffering from gout may present little or no evidence, so far as the urinary secretion is concerned, of any great acidity, and yet the uric acid may be easily obtained in a crystalline form from his blood.

There seems to be some little remaining doubt in the minds of scientific observers as to the true nature of the acid present in acute rheumatism; but general considerations, as well as actual chemical researches, prove that it is not the acid of gout.

There can be no doubt that this circumstance, as it points to a different origin in each disorder, also in great measure explains the difference of their progress and complications. Not that I am disposed to attach very much importance to the acid itself, either as a chemical or vital agent. It is quite possible that chemists may some day be able to prove that gouty acid is free from those peculiar properties which make rheumatic acid so liable, when present in excess, to set up cardiac inflammation. As yet we have no knowledge of the kind, and it is much more consonant with the present position of pathology, to recognise the two phenomena of acid formation and cardiac inflammation as common results of the same abnormal condition of the blood, than to assume that they bear any closer relation to each other.

There certainly is no *à priori* reason why in the one disease the inflammatory action accompanying the acid formation, should tend to reproduce itself in the heart when it does not do so in the other. Neither is it very apparent why so many of the more ordinary forms of disease should be modified by a gouty constitution, while in rheumatism, nothing remains of the attack except the damage done by inflammatory action. The subsequent heart disease has no specific character to indicate its origin, and in its progress exhibits no trace of special or constitutional tendency. If it be assumed that the analogy between the structures of the heart liable to rheumatic inflammation, and those of the joints is sufficient to explain its transmission to the central organ, it ought to be shown that in gout the structures involved are different, but no one has ever attempted to prove such a conclusion.

If this question be viewed merely with reference to the teaching of experience, it is found that, especially in a first attack, the two diseases present well-marked features which distinguish them most readily from each other. In acute rheumatism joint after joint suffers till perhaps every limb is implicated, and the patient is reduced to utter helplessness; and though the development of acute symptoms in one extremity is usually the signal for their partial subsidence in others, yet the number of joints simultaneously affected is, at times, very great. In gout, again, the first attack, in a large proportion of instances, commences with one great toe, and if it pass to the other foot at all, the whole force of the disorder seems to be con- centrated on the new point of attack, and very speedily the inflammation entirely subsides in the joint first seized. The local affection is much more intense, but its extent is also very much more limited, and the febrile disturbance is so slight that gout, however acute, could never properly be called " fever," in the same sense in which the word is so constantly applied to acute rheumatism. The evidence of con- stitutional disturbance is rather to be traced in an antecedent condition of general ill-health, which is so much more constant in gout than in rheumatism.

Again, we observe that gout is usually an affection of advanced years. When the hereditary tendency is very strongly de- veloped, the first attack may occur at a com- paratively early period ; but such cases are ex- ceptional, and are never quite so early in their appearance as cases of rheumatic fever, which

are very common before middle life, and are
frequently met with in childhood. This circumstance derives some significance from the
fact that the liability to cardiac complication increases immeasurably in proportion to the early
age at which rheumatism has first occurred.

Another very striking contrast is to be found
in the absence of any hereditary character in
rheumatism. There is no evidence to show
that a tendency to any form of the disorder is
ever transmitted to the offspring, by parents
who have been themselves great sufferers.
I conceive that an explanation of this circumstance may be indistinctly traced in the
difference in blood-change, which precedes the
first outbreak in each. It is more distinctly
seen in their relative liability to recurrence, as
the one becomes at once a constitutional disorder, while the other seems to be wholly
transient, and entirely to pass away with the
febrile disturbance which marks its presence.
A patient who has passed through one attack
of acute rheumatism may never be again subject to it. At all events it is not a disease
which shows any great tendency to recur.
Undoubtedly, if the circumstances under which
the disorder first appeared should be again

present, it may be assumed that the patient, in so far as he retains his individuality, would be more likely to have a second attack than one who, under similar circumstances, exhibited no such tendency; but this is all that can be said of the liability. With gout the case is very different. The first attack is but one of a series which lasts through life, returning at less and less distant intervals, until a period is probably reached when the patient is never wholly free. Unless the judgment of practical men greatly errs, there are many individuals who have a gouty tendency, whose every ailment seems to lean in that direction, who yet pass through life without one fit of the gout. If it had ever been developed, experience seems to show that it must have recurred.

Intercurrent diseases do not derive any particular characters from occurring in a patient who has once had acute rheumatism, as they so often do in the gouty diathesis. This observation applies especially to the febrile disorder which tends to originate cardiac disease. Mere muscular rheumatism is an affection to which some are constantly liable from almost every variety of cause : exposure to cold, acidity of stomach, or the depressed vitality of mental or bodily exhaustion, may each in turn excite the painful symptoms, but no febrile excitement marks their course. The same remarks apply, with even more force, to some chronic forms, in which the patient may be, after a time, never free from sensations of pain and stiffness of joints. Such conditions, however, are never the cause of disease of the heart.

It is essentially in the acute form, and most particularly when the pulse is full and hard, the tongue coated, and the whole surface of the body bedewed with a scanty, oily, ill-smelling secretion, that the chances of inflammation of the heart are greatest. When the pulse is softer, when the tongue is less loaded, when the skin is covered with more copious secretion, and still more when that secretion is most free from odour and most like the ordinary diaphoresis which follows a sharp febrile attack, then the chances of the heart becoming affected begin to diminish, and they diminish rapidly in proportion as fewer joints are implicated, still more rapidly as age advances. No fact is more clearly brought out by the statistics of rheumatic fever than this one of the influence of age. It over-rides almost all other concomitant circumstances, so

that while a child, left to nature, has scarcely
a chance of escape, though there be but the
smallest amount of joint affection, a young
adult may be considered perfectly safe so long
as one or two joints only are implicated,
however sharp the pain, however great the
heat, and redness, and swelling; and further,
a person advanced beyond mid-age may pass
through a very severe attack of acute rheumatism, with little or no risk of heart disease,
even if left wholly without treatment. In all
this we seem to see a close connexion between
the intensity of the fever and the liability to
inflammation of the heart, modified only by
the age of the patient. But there is something more than this. It is true that cardiac
complications seldom arise in the course of
such attacks of acute rheumatism as might
be mistaken for acute gout, considering the
age of the patient at the first seizure, the number of joints implicated, and the character
of the febrile symptoms present. Exceptions,
however, are not very uncommon, although
perhaps not very numerous. At a period
when I was endeavouring to collect statistical
information on this subject,* I not only had

* 'Med. Chir. Trans.,' vol. xxxv. p. 14, 15.

the opportunity of observing such cases for myself, but was also led to believe that they had been frequently noticed by others, who did not attach so much importance as I was in the habit of doing to the amount of fever in each case. It is also to be borne in mind, that though gout is especially the disease of advanced life, there are numerous examples of its inroads dating from a comparatively early period. Any one who has seen such cases, or who has perused their history, as recorded by others, must be convinced that some among them have exhibited such a degree of severity, in proportion to the age of the patient, as must almost of necessity have been accompanied by inflammation of the heart, if the disease had been of rheumatic in place of gouty origin. The records of medicine do not produce any authentic instance of endo- or pericarditis supervening during a first attack of acute gout, and we are thus bound to regard the immunity as dependent on something more than relative age and intensity when contrasting the effects of the two diseases.

In this point of view it is remarkable that a tendency much slighter in degree, but still unquestionable, to inflammation of the pericardium, and perhaps, too, of the endocardial membrane, has been traced in certain cases of Bright's disease. The late Dr. Todd used to refer to one form of this disease under the name of the " gouty kidney," although it has no special pathology of its own. Modern investigations have helped to clear up many of the obscurities which were at first left around a subject at once so difficult and of such recent discovery, but much still remains to be done. We have learned to discriminate among the various degenerations one or two very distinct forms ; but the granular kidney with which gouty symptoms are associated during life, does not as yet take any different place in pathology, from other examples of granular degeneration. In such cases gouty deposits are sometimes seen in the kidney itself, but they do not appear to be essential.

I do not know that the association of the two disorders has been in any way proved. It holds the same position in this respect as very much of our medical knowledge—having been asserted by one who was much looked up to as a teacher by those who had the benefit of his instruction, and meeting with a ready response

in the minds of all who have studied the subject; not indeed in the full sense in which he
employed the term, but in the modified ac-
ceptation of an implied association.   Most
persons who have had any considerable oppor-
tunities of observation have seen such cases.
They must have been in the habit of finding
albumen present in gouty subjects, but it is
not a frank, genuine gout which haunts the
patient.   He may have the marks of gouty
deposit, and may be able to refer to very sharp
and severe attacks in days gone by, but the
cases which suggest the probability of kidney
disease to the mind of the observer, are those
lingering ill-developed forms, in which it is
very difficult to say whether the pain and stiffness are due to rheumatism or to gout ; and
the question is only settled by the knowledge
that the previous attacks have been gout, and
the consequent probability that the present illness is gout too.   Such considerations have so
frequently come under my notice, that the
examination of the urine is rarely omitted
when cases of indistinct chronic gout come
under treatment.   I feel no doubt that if the
materials for making the calculation were
brought together, it could be shown numeri-

I

cally that the relationship is not merely accidental.

The time has not yet arrived for settling
the further question—viz., which of the two
diseases has the first place in the order of
causation. Are we to suppose that the gouty
symptoms are dependent on the obstruction
to the excretory function of the kidney, or
does the excessive formation of uric acid in the
system excite disease of the kidney by its
elimination through that organ? Difficulties
seem to stand in the way of an answer in the
affirmative to either suggestion. If, on the
one hand, it be alleged that the suppression of
the elimination of uric acid is the cause of the
gouty symptoms, they ought to be present in
every case in which this event occurs. And,
on the other hand, if a constant excess of uric
acid in the blood be supposed to provoke the
kidney disease by its transit through that
organ, then all gouty persons should have
albuminuria, which we know they have not.
It seems clear that whatever be the relation
between the two diseases, they are not necessarily associated together, and we must look
further back in the order of causation to trace
out the association.

It has been already noticed that general experience points to distinct forms of disease as resulting from the abuse of the various kinds of alcoholic stimulants, and among these we recognise both gout and albuminuria. It is believed by some observers that the granular kidney is the ultimate result of a slow process of inflammation, probably excited by a morbid or poisoned condition of the blood. This would seem to be the prevailing idea in Dr. Johnson's assertion of a "chronic desquamative nephritis." To others, again, the facts seem better explained by the hypothesis of a degeneration of tissue. One of the latest exponents of this view regards any injury done to the secreting apparatus as secondary to, and consequent upon, the degeneration of the interstitial substance. This hypothesis classes it with other forms of degeneration as a process of mal-assimilation. It seems superfluous to observe, that each hypothesis alike refers the granular kidney to an altered condition of blood. I have ventured, in an earlier part of this volume, to suggest the possibility of a change in the blood-globules as the main cause of gout ; and I am very much disposed to regard all the degenerative processes, wherever occurring, as being, in like manner, dependent on some more enduring alteration of the blood, than we can suppose ever takes place in its serous and soluble constituents. It is unnecessary, however, to determine in what particular respect the blood deviates from its condition in health. Under any hypothesis that may be suggested, we can never lose sight of the circumstance that blood-change is an essential element in their causation.

We thus get one step nearer to an understanding of the coincidence which links the two diseases together. It is quite intelligible that both forms of disease should re-act upon each other—that the same tendency which causes the retention of the uric acid in the blood in an attack of gout, and interferes with its elimination by the kidney, should, at the same time, lay the foundation of subsequent degeneration ; and that the suppression of the solid constituents of the secretion in chronic cases of albuminuria, should oppose the outlet to the excessive uric acid formation which takes place in gouty subjects. We can thus understand that acute gout may be almost the starting-point of granular degeneration, while, in chronic cases, the kidney disease helps to maintain, if not to originate, an almost constant condition of chronic gout. But we do not stop here. By going a step further back in the history of each, we trace in the altered blood a circumstance sufficient to account for the development of each form of disease. It is only in their after progress that we see the two diseases re-acting upon each other, and it is in this modified sense alone that I should be disposed to subscribe to Dr. Todd's assertion of the existence of a gouty kidney.

Dr. Garrod, in alluding to this subject, merely says that, of the number of cases quoted by him, a large number had albuminous urine. It has never been made a subject of special enquiry by pathologists, and the only one who has referred to the statistics of the subject dismisses it very briefly with the simple statement that the facts indicate some relation of cause and effect.*

More trouble has been bestowed on the proof of the relationship of disease of the kidney to inflammation of serous membranes. In none perhaps is this more marked than in the pericardium; and this condition is constantly associated with a deposit of recent lymph

* 'Med. Chir. Trans.,' vol. xliv. p. 172.

upon the valves of the heart, seen on *post-mortem* examination even when it has not been possible to trace any disease of the heart during life. That such an event is possible, or even probable, in gouty subjects, we are not prepared to deny; but it is quite certain that inflammatory action is not the principal cause of disease of the heart associated with gout. The valvular lesions met with are much more commonly the result of degenerative processes on which inflammation has been more or less manifestly engrafted.

CHAPTER VI.

DISEASE OF THE HEART IN GOUTY SUBJECTS.

————◆◇◆————

Its threefold relation to Gout — Origin in inflammation or degeneration — Causes of inflammation — Apparently independent of Gout — Degeneration often associated with Gout — Fatty degeneration — Atheroma — Also occur independently of Gout.

IN gouty affections of the heart three orders of cases seem to present themselves, and are marked out by very different circumstances in their history and progress. First we have to distinguish those in which heart disease has sprung from some wholly independent cause in a person who is subsequently attacked by gout. Secondly, a large number of cases may be met with where the mal-nutrition accompanying a gouty diathesis, slowly developes disease of the heart; and Thirdly, we recognise cases of metastasis to the heart, the organ being either previously healthy, or, as is more commonly the case, affected by some organic lesion.

In one point of view these three classes have a mutual dependence on each other, as a weakened condition of the organ predisposes it to metastasis, and renders the complications more serious and more apt to terminate fatally than when the heart is healthy. The diseases on which gout may be engrafted are simply those of inflammation or degeneration with which we are so conversant in every-day practice. They present no features by which they may be distinguished from others which are free from this complication. A gouty person, however, may be said to be less exposed to the chances of inflammation of the heart than one who has not inherited that tendency. He escapes the great cause of such attacks by the acid-forming vice in his constitution tending to the formation of uric acid in excess, and not the acid of rheumatic fever. Gouty patients are not rheumatic. A hybrid disorder was, as has been stated, at one time assumed under the name of rheumatic gout, but more recent researches prove that the two do not go together. We must look to more correct diagnosis as the only means of discriminating such cases, and giving us more correct ideas of their progress and termination. But we

cannot overlook the fact that certain forms
of chronic rheumatism and chronic gout bear
a very close resemblance to each other, and
have many points of contact. Were I to
venture to cite my own experience on such a
subject I might affirm that, when the two bear
a close resemblance to each other, the concomitant diseases are also closely allied. The
same form of disease of the kidney is constantly
seen in both orders of cases, and so, too, we
find the same general condition of the heart;
the disease in each being probably caused by
some form of mal-nutrition.

Inflammatory disease of the heart may,
however, be traced to the existence of other
inflammations in the chest. Pericarditis is
frequently associated with pleurisy, and when
not caused by acute rheumatism it seems
reasonable to assume that the inflammation
of the pericardium, as the more rare event, is
excited by the inflammation of the pleura,
which is such a frequent result of exposure to
cold. From the existence of pericarditis again,
a long train of cardiac diseases follows as an
almost necessary consequence. It has been
ascertained that the attack may be arrested so
as to leave no trace of its existence beyond the

presence of a white patch on the surface of
the membrane, as was first pointed out by Mr.
Paget; * but when the disease passes on to
the formation of adhesions, the muscular structure sooner or later becomes altered in texture;
and when the valvular apparatus is also involved the damage may be regarded as
irreparable.

It has been just remarked that inflammation of the heart is not due to acute gout. And
even when an association can be traced through
the presence of albumen in the urine, it is not
during a gouty paroxysm that the inflammation begins. Hence we are justified in affirm-
ing that, when the proofs of its previous existence are most distinct in after years, the gout
is merely a coincident disorder. The fact,
when ascertained, derives its chief importance
from the circumstance that valves once damaged
by inflammation are always more ready to
become the seat of diseased action than they
were before. They more readily receive a fresh
deposit of fibrin, and are also more liable to
become atheromatous. For the same reason,
too, they may become the site of a deposit of
urate of soda, which probably never occurs in a

* 'Med. Chir. Trans.,' vol. xxiii. p. 29.

perfectly healthy valve. My own conviction, from the investigations I have made on this subject, is, that it is only found when the valves are already damaged by previous disease. It does not, therefore, seem to be consonant with experience to suppose that inflammation of the heart is ever modified by an undeveloped tendency to gout lurking in the constitution. We shall have to consider presently the influence of the gouty paroxysm upon a damaged heart, the question now under consideration is, how far gout which has not yet been developed, but only exists if present at all, as a remote tendency or diathesis, can affect cardiac inflammation. The probability is, I think, all in favour of the assumption that, as the deposit of urate of soda does not take place, the inflammation is not much modified in its course and progress at such an early period in the history of gout. After its effects begin to be traced elsewhere, and the disease is no longer an abstraction, just as we find it modifying inflammations in other organs, so is it probable that it would modify cardiac inflammation. As yet we know but little of either endo- or pericarditis apart from their occurrence in connection with acute rheumatism ; their clinical history is scanty and obscure, and the observations that have been made are chiefly derived from *post-mortem* appearances. It is not very surprising that, under such circumstances, we are left so much to conjecture, especially when it is considered that the whole number of cases must be exceedingly small in which either form of inflammation had occurred in a gouty person.

When we turn to degeneration as a cause of disease of the heart, we find that it is not at all unusual in gouty persons ; and general considerations lead to the supposition that the gouty diathesis is one of the direct causes of its development. It was long assumed that chronic inflammation of some sort or other was the fundamental process in degenerative disease. It was hypothetically assumed that in such states as cirrhosis of the liver and granular kidney, there was first an effusion of fibrin, and that in the course of its absorption the natural tissue was removed, so that the organ ultimately presented a shrunken or dwindled appearance. Recently acquired pathological knowledge leads us to doubt whether such a course is ever followed in such cases. It associates together in the same group the athe-

romatous condition of the lining membrane of
the heart, the fatty degeneration of its muscular
walls, and the granular and lardaceous conditions of liver and kidney, as being all common
results of mal-nutrition. Ultimately, no doubt,
these will be separated from each other so as
to form smaller sections, either in relation to
their causation or their microscopical elements.
At present we must be content to regard them
as simply due to imperfect preparation, and
assimilation of the elements of healthy cell-
growth. At the same time it is necessary to
exclude from consideration those special types
of morbid growth which develope into cancer-
cell or tubercle, and which are never found in
the heart without their being widely distributed
over other organs of the body. The less
defined forms of mal-nutrition arise from a
variety of causes, but they have one common
character, viz., that they originate a condition of blood, which does not afford a healthy
pabulum for the nutritive forces to lay hold
of and assimilate.

One of the most remarkable of these changes
is that which is known as fatty degeneration.
In the advanced stage, the striated appearance
of the muscular structure is lost, and the fibres

present under the microscope the aspect of a
tube filled with granular matter and oil globules. A heart which has undergone this change
looks yellowish to the eye, and has lost its
consistency. We are not fully conversant with
the various stages of this process, but it is
believed that though the fat accumulates apparently within the sarcolemma, it is not by a
process of transformation so much as by one of
substitution. The muscular fibre first becomes
atrophied, and then oily matter takes the place
of the contractile tissue which has been removed. It is impossible to overlook the importance of such a process. The atrophy of
the fibre must of necessity weaken its power.
When there is any impediment to the flow of
the blood, the walls yield and dilatation of the
cavities takes place ; in other circumstances
the whole organ dwindles and wastes. The
more it increases in size without a corresponding increase in thickness, the greater in quantity
is the adventitious deposit of fatty matter, and
the more marked do the symptoms of a weak
heart become. The obstruction which first
acts by causing an accumulation of blood, helps
to produce laboured and irregular action, with
all its attendant consequences. If the current

of the blood be not obstructed, and simple
atrophy takes place, whether with or without
fatty deposit, the distress of the patient is
seldom so great till some occasion arises when
a demand is made on the heart to propel the
blood current with unusual force or rapidity,
and then its power suddenly gives way. In
neither case is there any evidence in the early
stage of what is about to take place. The valvular disease when present may be easily traced, the
progress of enlargement may even be accurately
followed, and yet the existence of fatty degeneration must be almost necessarily a matter of
inference. In the absence of any direct impediment to the flow of blood, it may seem
somewhat more easily proved that the heart's
walls are unequal to the requirements of the
circulation, but even then it does not follow
that a feeble heart is necessarily a fatty one,
and want of power is all that can be asserted regarding it, with any degree of certainty.

Valvular degeneration, like that of the
muscular structure, is very often found in
association with gout. Its progress seems
wholly distinct from inflammation, and is
marked by the same characters, which can be

traced in the diseases to which the lining
membrane of the arteries is subject. This may
be regarded as the necessary result of the
laws of growth, as the lining membrane is
continuous from the interior of the heart to
the ultimate branches of the arterial tubes,
where they terminate in the capillary vessels.
The change of structure which then takes place,
explains why the same form of degeneration
is unknown in the veins. The patches of
atheroma, which are seen on the mitral and
aortic valves, and are occasionally, though
much less frequently, traced on the lining
membrane of the heart, become abundant and
of considerable extent on the aorta, and are
constantly found in the smaller arteries. Their
comparative absence on the lining membrane
of the heart is probably explained by the circumstance that it is there spread out upon
fleshy fibre and not upon elastic tissue, which
forms the preponderating material in the coats
of the arteries, and alone constitutes the flaps of
the valvular apparatus. It is not in the first
departure from health that the danger to life
consists. A valve is scarcely less efficient
because it presents the appearance after death
of a few patches of atheroma, nor does the

artery as yet lose much of its resiliency. But further change speedily follows. Atheromatous deposit gradually produces thickening and condensation of the underlying tissue, and very soon takes on, under circumstances favouring its development, a further change into calcareous deposit. And now the valves become rough and rigid, either obstructing the onward current of the blood, or permitting its regurgitation. The arteries at the base of the brain are apt, in like manner, to become hard, unyielding, and friable, and apoplectic hemorrhage is the result. In other parts of the body they may similarly give way, when an aneurism will form, or they may offer such an impediment to the circulation that senile gangrene follows.

These changes are, so far as we know, the simple results of mal-nutrition. They are by no means necessarily associated with gout, and are so only because in the gouty subject the blood is unhealthy, the due assimilation of nutriment is interfered with, an unusual amount of uric acid exists in the body, and substances which are perfectly innocent as the food of a healthy man, and are absolutely necessary to prolong life in one suffering under other

K

forms of disease, become injurious to one who has a gouty tendency.

If it be still an open question whether the degenerations met with in gouty persons have any closer relation to the so-called diathesis, than is implied when they are both ascribed to mal-nutrition, it must be remembered that the tendency to gout is an hereditary one, while the degenerations are certainly not proved to be so. Gout is distinguished by a broad line of demarcation from other forms of imperfect nutrition, but the degenerations of the heart are found alike in association with it, and with diseases which claim an independent origin. I am rather inclined to the belief that the process is really quite distinct, and that if the imperfect nutrition which accompanies gout in the form of constant dyspepsia, or granular degeneration of the kidney, tends to its development, it is just as constantly met with in cases where there is no gouty tendency.

CHAPTER VII.

THE GOUTY PAROXYSM AS IT AFFECTS THE HEART.

Erratic character of Gout — A blood disease — With inflammation of Joints — Analogy of Pyæmia — Transitory character of
the local affection — No true Metastasis — Blood-poison —
False views arising out of this term — Mal-assimilation in relation to Gout — Its resemblance to Delirium Tremens — The
blood loaded with effete matters — Irregular action of the
heart in Gout — Its relation to fatty degeneration and
atheroma.

WE have next to consider the effect of gout
upon the heart, whether previously altered by
disease or not, under those circumstances to
which the term metastasis has been so frequently applied. The irregular and retro-
cedent forms of the disease when the joint
affection suddenly subsides, and its severity is
only traceable in its effect on some internal
organ, are among the most interesting subjects
of study in the whole range of practical medicine.
The heart, the stomach, the brain, may each
become the seat of the disorder, and the immi-

K 2

nence of the danger is only measured by the importance of the organ implicated. But before we admit the existence of a true metastasis, which is scarcely ever affirmed regarding any other disease, let us enquire what the term really means. It is assumed that the disease retreats from the part first attacked and locates itself somewhere else, and generally in some internal organ. Now it must be borne in mind that though gout in its first manifestation very often limits itself to one joint, and when this is the case, it is most commonly the first joint of one great toe, yet it is still essentially erratic. The inflammation of the joint, as has been shown by Dr. Garrod, seems to be closely associated with an arrest at that particular spot of the uric acid present in excess in the blood. Why this is so we do not know, but it is believed that the attack is invariably accompanied by an effusion of urate of soda in the part. In a large proportion of subsequent attacks of gout, and in most of those in which it is not limited to the great toe, the inflammation appears and disappears in several joints in succession—sometimes returning to the spot where it first exhibited itself—sometimes leaving it entirely and

remaining fixed in some other joint. This peculiar action of the disease has been rather hastily
assumed to be an attempt at eliminating the
poison. I think there is no evidence to bear
out this conclusion beyond the circumstance
that the effusion which causes the swelling of
the part is always accompanied by the deposit.
But if uric acid be contained in excess in the
blood and held in solution in the serum, it
would seem to be impossible that serum
should exude without carrying the urate of
soda with it. If a joint which has been the
seat of rheumatic inflammation be opened at
an early period it is found occupied by a fibrinous exudation, the fibrin of the blood being
in large excess in rheumatic fever. Just as
well might this be assumed to be a process of
elimination of the poison as when urate of soda
is found in the joint after an attack of gout.

When inflammation is excited by an irritant,
certain stages may be traced both by microscopical examination, as shown by Mr. Whar-
ton Jones, and also by the more evident and
palpable effects with which every one is familiar. For example, in the application of a
sinapism for a very short time, there is seen
a certain amount of redness of the skin, which

will last only an hour or two ; by longer application the redness may be made to last from
twelve hours to three or four days ; still longer
irritation by the mustard will produce vesication, which leaves traces of its existence for a
much longer period, and suppuration may follow, which requires perhaps months ere the
skin be restored to a perfectly natural appearance and condition. If this simple illustration
be applied to gout, it is easy to see how the
redness and tumefaction of the skin may disap-
pear in a few hours, or may last for some days.
Vesication indeed never follows, and when the
cuticle peels off, as it not unfrequently does, it
is owing to an arrest of healthy nutrition,
whereby a breach of continuity occurs in the constantly progressing elaboration of the cuticle,
just as an injury to the matrix of the nail
arrests for a time its development; when
the process of growth recommences, the old
nail does not retain its connexion with the
new one.

The joint affection in gout is manifestly not
due to any external agency. It may be excited in a person predisposed to the complaint
by a blow or a strain, but much more frequently the patient goes to bed at night and

is awaked, after a few hours' sleep, by his
tormentor. And when the evidence of some
external cause is most distinct, we are still
able to satisfy ourselves that the agency is
not sufficient of itself to set up an inflammation attended with so much heat, redness, and
tenderness, and to feel perfectly certain that it
could not have excited it in a perfectly healthy
constitution. The irritant must then be internal, and there can be no doubt that it is in the
blood; but the circumstance is as yet unknown
to us which locates it now in one joint, now in
another. Be that cause what it may, we know
by experience that it operates sometimes on
one joint, sometimes on several, attacking them
either simultaneously or in succession. The
inflammatory action, when it does not pass a
certain stage, leaves the joint to all appearance
just what it was before the attack in a very
few hours after the irritant ceases to act
upon it.

A very analogous process is seen in the disease known as pyæmia. Here, again, there is
some material circulating in the blood which
tends to produce local disease in various parts
of the body, which are attacked either together
or at successive periods: inflammatory action is

set up, and the tendency of that inflammation
is to the formation of pus. When pus is really
formed, a considerable period must of course
elapse before there can be a return to a state
of health, even under the most favourable circumstances; yet instances constantly occur in
which some one or more of the parts attacked
never pass beyond the first stage, and all trace
of inflammation subsides in a very few hours. I
do not know that any theory has fully explained
why pyæmia attacks one organ of the body
more than another; but there must be some
localising influence that determines it to the
skin in one person, to the liver or the lungs
in another. Here the analogy seems almost
perfect, and may perhaps help in suggesting
an hypothesis which will serve to guide our
views to some extent in considering the transi-
tions of gout. Let us not, however, carry it
too far: the collections of pus once formed in a
case of pyæmia are very different from the
gradual accumulations of urate of soda; and
the one disorder attaches itself to its victim
for life and descends to his posterity, while the
other, when subdued, quits its hold altogether.
But just as in pyæmic cases, patients may
escape the pus-formation in some of the in-

flamed centres; so it seems highly probable that sometimes in gout the attack may be so transient as to leave no trace of chalky deposit.

The two points just alluded to seem to afford a sufficient explanation of its erratic character. The arrest of the morbid material in various joints, localised, as has been said, by some circumstance as yet unknown, occurs in a more or less rapid succession; and this local affection is often of a most transient character, leaving possibly no relic of its past existence. If these two considerations receive their due weight, we shall have no difficulty in accounting for the rapid transference of its attacks without the necessity of assuming that its subsidence in one part is necessarily connected with its appearance in another. And if, like pyæmia, it has a tendency also to fix itself on internal organs, it is here that the correspondence of the two affections is least seen, and their difference most shows itself. Gout does not so readily attack internal organs in an acute form, and when it does so its duration is more transient than under any other circumstance. The very reverse holds true with pyæmia. The one disease presents in most cases a de-

cidedly erratic character, the other is more
constantly fixed.

The term metastasis implies a theory which
I believe to be at variance with these views.
It seems to mean a transference of the *materies
morbi* from one spot to another; not as I have
sought to explain it, an arrest of the altered
blood in some new situation, the former attack
ceasing less as a consequence of the new seizure, than simply because the action had not
gone beyond a certain stage. The view expressed in the term metastasis cannot be récon-
ciled with what we know of the real pathology
of gout. The local affection is but a manifestation of an altered state of system, which
probably has for a long period given rise to
symptoms of dyspepsia and disordered assimilation, and these are now known to be coincident
with an excess of uric acid in the blood,—itself
only another symptom of the gouty condition
of the patient. All are to be regarded alike,
and though the joint-affection is a much more
striking phenomenon than the others—and its
presence seems to convey a more definite idea
of the disorder, both to patient and physician—
yet it must never be forgotten that it is in its
nature a secondary, and not a primary affec-

tion. It is very probably true that the intensity of the action in one part has some
influence in modifying it in another; and no
doubt there is a great amount of reason in
the view which advocates the solicitation of
gout in the foot to relieve the oppression
of internal organs. But this is no more
than is seen in many other forms of disease.
An attack of pneumonia will very often arrest
the progress of some minor disorder, and that
again may be relieved by means of counter-
irritation. We do not employ what used to
be called revulsives so much as was the practice in former days; but this is only because
we have attained clearer ideas of the causes
of disease, and we seek more constantly to
alter the state of system on which their presence depends, and are less concerned with
the local malady. When nothing more was
known of gout than the joint-affection, it
was very natural to suppose that if an internal
organ was affected, while the joint simultaneously seemed to return to a state of health, the
disease had departed as an entity from it, and
had fixed itself on the internal organ—probably carried thither by the blood. It is
needless to say that such hypotheses have no

foundation in fact, but the language in which
they were couched prevails to this day, and
tends in its constant employment to lead the
mind away from the realities with which alone
medicine ought to deal. Disease does not consist in anything superadded to the structures,
and it would be well if we could abandon the
phraseology that speaks of it in this manner.

The very same objection applies to the loose
way in which modern writers speak of what
are called blood-poisons. This seems to me
to be only an adaptation of the old theory to
the requirements of modern science, and
sways the views of those who employ the
term perhaps more than they are themselves
aware of. It is quite true that in those forms
of disease to which the term is applied there
is, in the first instance, some morbid material
taken into the body which has a deleterious
effect on the constitution, just as when a man
swallows a poisonous quantity of any vegetable
or mineral substance. But what follows? A
period of incubation occurs, in which there is
no appearance of the health being deteriorated.
It is impossible to tell whether the poison will
pass out, and leave him just where he was,
or will ultimately excite a series of changes

which constitute disease. To this stage alone
is the term poison properly applicable; it is
during this period alone that it can be eliminated; and if eliminated there is no disease.
When disease has been set up a new series
of actions begin to manifest themselves; and
one effect of these changes in certain forms
of blood-poisoning is that they produce a
morbid matter of some kind or other, which
does not much influence the subject of the
change, but when given off will, in a number
of persons exposed to their influence, excite
similar actions. This tends to keep up the
false impression that the poison still influences
the patient himself.

Many other substances besides animal poi-
sons will set up disease in the body, but they
are not self-generating, and therefore do not
suggest the same fallacy. For example, a
corrosive poison may kill by the intensity of
its immediate action, and to this we carefully
restrict the term poisoning as produced by
such substances. But it will also cause the
destruction of a certain amount of tissue, and
so set up a series of secondary actions such
as inflammation, suppuration, and subsequent
adhesion of parts of the fauces, esophagus, or

stomach: and when these secondary effects are produced we no longer think or speak of the poison as being lodged in the system, we do not seek to eliminate it. A more notable example is to be found in alcoholic poisoning. When a man has swallowed such a quantity of alcohol as to stupefy him, he is to all intents and purposes poisoned by the alcohol, whether it kills him outright, or the poison becomes eliminated and he recovers consciousness. But if he go on from day to day swallowing alcohol in excess, ultimately diseased action arises in the form, for example, of delirium tremens. He is now no longer poisoned by the alcohol which is in his blood, because the whole or the greater part of it has been already eliminated; but his tissues are altered, disease has been set up, and very frequently the administration of alcohol in some form or other is essential to his recovery. In such a case the morbid material given off is not infectious, and therefore the analogy is not perfect. Still I conceive that it approaches much more nearly to the condition of the body when labouring under those diseases which owe their origin to infection, than is generally assumed when blood-poisons are spoken of, and treatment is recommended for the express purpose of eliminating the poison. The patient does not suffer because his tissues are generating a poison which may produce infection in another, but the poison is generated because the tissues are in a state of change.

I have been led into this digression because the very same language is employed in speaking of gout. As a consequence of this confusion of language, we now find gout classed along with scarlet-fever and small-pox as a blood-poison,* although the circumstances attending it are so entirely different. All the symptoms of gout prove that some alteration has taken place in the condition of the blood. It is no longer in such a state that it can circulate freely through the body without exciting abnormal actions in certain tissues, and along with this there is unquestionably an excess of uric acid. But we shall very much mistake the nature of the disease, I apprehend, if we regard this excess as the poison of gout, or if we suppose that when we have found a means of eliminating the uric acid, we shall eliminate gout. No one will for a moment assume that gout is, in the proper sense of the term, a blood-

* See the Registrar General's Returns, *passim.*

poison. It does not spring from the absorption of any infectious material. Gradually and slowly the perversion of the functions of the assimilating organs alters the condition of the nutritive fluid, and possibly, too, of the solid structures; and then suddenly a twinge is felt, probably in the great toe, and regular gout sets in. It is in this way alone that any explanation can be given of its hereditary character. The child inherits from its parents an assimilating apparatus weak and imperfect, and liable to produce that form of deteriorated blood which we only know in its fully developed condition. If placed in very favourable conditions by careful diet and an active mode of life, the blood may retain its purity through life ; if the circumstances be changed, either by carelessness or necessity, the blood changes its character too.

When, on the other hand, gout is acquired, it is only because the circumstances are more than usually inauspicious, and the assimilating organs, though in the first instance free from any inherent vice, become unfit to discharge their functions in a healthy manner. It is not a little remarkable that when they have once reached this condition they are always liable to the recurrence of the same functional derangement; for gout very seldom limits itself to one attack. To my own mind the analogy between gout and delirium tremens is very close. We have not indeed the means of knowing whether delirium tremens ever becomes hereditary. The probability is against such a supposition, for the simple reason that it has not been noticed; but it is quite possible that the child of a parent who has suffered from it may acquire it more readily than another. It is a hideous and revolting disease, and the circumstances which give rise to it are so contrary to all the best feelings of humanity that we are entitled to hope that such an observation may never be made. With this exception, however, I think if we analyse the two disorders we may discover much of resemblance with reference to the question of blood-poisoning between them.

In the early history of delirium tremens, we trace the constant introduction into the system of an amount of alcohol which the constitution of the patient cannot readily bear. At first, health and activity enable him to eliminate the poison—for alcohol is a poison when taken in excess. By-and-bye the assimi-

lating organs begin to suffer, dyspeptic symptoms indicate their inability to convert the
food into healthy nutritious blood, the brain
structures, too, are probably affected by the
constant irritation of blood charged with
alcohol, and are imperfectly renovated because
the nutrition is generally interfered with;
trouble and anxiety of mind may perhaps
be added, and then sleep is first disturbed,
next the "horrors" begin, and at length the
disease is fully developed. In some cases the
continued employment of the stimulant after
its commencement may vary some of the principal features of the disease; but it is quite
certain that no alcohol need be present in the
blood for its development, and some authors
even assert that it commonly begins in consequence of unusual and forced abstinence.*

In acquired gout we shall find a history
very analogous. The same constant employment of some stimulant, which acts as a poison
only because taken in excess, is followed in
due time by perverted assimilation; dyspeptic
symptoms supervene and lead to the formation of unhealthy blood. Some cause of mental
anxiety follows, or sedentary employment is

* Watson's 'Practice of Physic,' Lectures 23 and 24.

coupled with much harassing business, and
after a time the changes which have been
long going on show themselves in a fit of the
gout.

The analogy is, of course, imperfect in many
particulars. The character of the stimulant
differs in either case, and the alcohol con-
sumed in ardent spirits, which is the usual
antecedent of delirium tremens, must necessarily excite the brain in a different manner
from those milder beverages which develope
the gouty tendencies. There is, too, a very
striking difference in liability to recurrence.
To eradicate any tendency to delirium tremens
a certain amount of moral control is alone
necessary: the patient debarred from his usual
excess is in no danger of a recurrence of the
disease. The case is far otherwise in gout.
No amount of restraint will wholly exempt one
who has at any time suffered from the disease
from the possibility of its recurrence. Still, on
the whole, there is a sufficient resemblance in
the course and causation of that imperfect
nutrition which leads to the development of
either disease, to afford a safe basis for argument on the question of blood-poison. And
the conclusion seems inevitable that in both

alike the term is one very liable to mislead, because it assumes the presence in the blood of some positive deleterious agent; in reality we know no more than that the blood has undergone such a change as renders it unfit to maintain the healthy activity of one or more organs of the body; and to these, consequently, the disease at first sight appears to be limited. The presence of the effete material found in the fluid is not a poison in the true sense of the term, but only an evidence of blood-change. Its importance is to be measured rather by the value of its indication as a symptom than by its influence in exciting secondary phenomena.

The conclusion thus forced upon us explains the dependence of cardiac affections in gouty persons on altered blood. They are chiefly marked by a sensation of great oppression at the præcordium, difficulty of breathing, and a sense of suffocation, with irregular action of the heart. This irregularity is often a very remarkable feature, as it comes on without any previous symptom of disease having been noted; and though it may pass off entirely, and the patient may resume the appearance of perfect health, yet it not unfrequently leaves the heart in a state of permanently enfeebled action. Its regularity never returns, and its intermissions are accompanied by occasional spasm, which in character is very closely analogous to angina pectoris. The patient may die in some such attack when of more than usual severity, or he may suffer from the ordinary train of symptoms which attend a weak heart, and life may ebb out in the secondary disorders set up by it. All this is very intelligible when it is remembered that in chronic gout certain changes of structure are often found, which are due to the morbid condition of the blood itself.

When a heart affected by gout is examined with the stethoscope, the sounds which are heard may either indicate traces of valvular disease, or merely enfeebled action; and it is perfectly well understood that both conditions depend upon imperfect nutrition, as it affects either the lining membrane of the heart or its muscular structure. It is a very remarkable circumstance, and one well known to medical men, that a patient may go on with a damaged heart for many years without his being in the least aware of the serious malady under which he is labouring. Suddenly and unexpectedly

to himself, in consequence of some unusual
strain, the action becomes tumultuous and
irregular, and no care or quiescence, no
remedial means of any kind, ever serves to
restore it to the normal rhythm which its
movements previously exhibited. It is vain
to speculate on this curious phenomenon. It
does not necessarily imply, though this, too,
may happen, that rupture of some fibre has
taken place. Such, at least, is not the history
of the cases now referred to. We speak simply
of those in which changes of long and slow
development have taken place in the muscular
structure, associated, probably, with valvular
lesion; the change of tissue is perhaps of
a compensating character, and life itself could
not have been maintained had no such
alteration taken place; the power of the
organ is, consequently, only sufficient to keep
up a regular circulation so long as no unusual demand is made upon it. But when an
extra strain comes, the action is not merely
quickened as in health, it is also irregular;
this irregularity further impedes a laboured
circulation, and the consequent stagnation of
blood in distant vessels keeps up the increased
demand on its power which it is unable to

meet, so that there can be no return to quiescent and regular action.

In gout the sequence is not very different. Here it is not to any sudden strain that the irregularity is primarily due, but to the presence of gout in the system: the effect however is the same. Irregular action is excited, and if it so happen that the heart has already undergone important changes, the irregularity interferes with the proper propulsion of the blood, partial congestions necessarily follow, and it is difficult, perhaps impossible, for the heart to resume its normal rhythm.

The difficulty, too, is much enhanced by the circumstance that both in the lungs and in the kidney changes due to gout have probably taken place, so that the pulmonary circulation on the right side of the heart, and the systemic circulation on the left side, are alike difficult, and partial stagnation in each cannot fail to occur.

The form which structural changes in the heart assume in long standing cases of gout greatly influences the result of such attacks. They are, as has been already stated, of the degenerative, not the inflammatory class. The muscle though enlarged, has lost rather than

gained in power when thus affected, and it is
in such cases that irregular action tends to
become persistent. It affords a pretty sure
indication that the muscular tissue has escaped,
when we find the pulse resume its regular force
and evenness, as soon as the gouty attack has
left the central organ and reappeared more
distinctly in the joint. So many causes may
contribute to inequality, if not to distinct
irregularity, of the pulse, that no one can pro-
nounce with certainty on the state of the
heart's walls without long and patient investigation and frequent observation. To some
persons intermittent action seems the normal
condition, and circumstances almost inappreciable to our rude means of research develope the
intermission to such a degree that it assumes
all the characters of irregularity. With the
change in rhythm comes almost certainly a
change also in power, and in the absence of any
valvular lesion, a hasty conclusion would probably assert the existence of some form of
degeneration or atrophy, when none such exists.
It is clearly not to gout alone that we must
look for the explanation of such occurrences.
But gout is just one of those circumstances
which must not be overlooked in their study,

though perhaps the danger is rather that it should be too constantly accused of their production than too generally ignored.

The valvular lesion in gouty subjects is usually degenerative, and in its early stages does not assume such importance with reference to enlargement of the heart as when it is the product of rheumatic fever. The murmur heard through the stethoscope probably does not prove that the valves are unfit to perform their functions, or indicate any serious obstacle in the way of the circulation, but it has a different significance which must not be lost sight of. If no history of an inflammatory origin can be obtained, we may be pretty certain, in gouty persons, that the lesion has begun by a deposit of atheroma; and as the lining membrane of the arteries is still more liable to be affected by this degeneration than that of the heart itself, the systolic murmur serves as an indication that in all probability it extends further through the circulatory system. This is indeed the only source from which we can derive a suggestion of its existence in the early stage, and its importance must be measured by what is known of its consequences. Loss of resiliency in the arteries from this cause places an obstacle in the way of the circulation of far greater importance than the slight roughness or induration of the valve which becomes perceptible by the aid of the stethoscope. A *bruit* of this character, however unimportant in itself, may indicate more peril to life than a much louder one which depends for its production on changes originating in endocarditis; the one affection being simply local, while the other involves more or less the whole of the arterial system.

CHAPTER VIII.

GENERAL VIEW OF DISEASE OF THE HEART.

The importance of enlargement — Comparatively harmless murmurs — Systolic aortic — Its consequences — Systolic mitral
leading to more serious disease — Errors in diagnosis regarding
the mitral valve — Aortic regurgitation — Enlargement its
invariable sequence — Theory of the murmur heard — Enlargement occasionally overlooked — Hypertrophy — Dilatation
— Their usual causes—Considered as an aid to diagnosis.

CHRONIC disease of the heart must be considered altogether apart from its relation to
gout. I do not propose here to enter into a de-
tailed examination of this most interesting subject; but simply to refer to one or two questions
in relation to it which seem to me of some importance. Valvular lesion may last for an
almost indefinite period without producing
any distress or giving rise to any symptoms by
which its presence may be indicated. This is
more true of certain forms of disease than of
others, and hence an early discrimination of
the character of the imperfection of the valve

becomes of no small importance. When we
inquire into the reason of the difference it will
be found to resolve itself into the greater liability of one form of disease to excite changes
in the muscular structure than another. To
any one conversant with the clinical history of
heart disease this relation is well known. It
is not till the heaving movement of the chest
at each pulsation indicates the presence of considerable hypertrophy, or the persistent irregularity of the pulse proves the muscular power to
be insufficient for the purpose for which it was
intended, that any inconvenience arises from
valvular lesion. A very distinct *bruit* may be
heard with the stethoscope, and the individual
may be on close questioning found to be not so
long-winded as others of his own age, or as he
himself was when the valves were free from
disease; and yet the patient may go on for
years without very material discomfort, and
even without a suspicion that there is anything
wrong.

The foremost place in the rank of comparatively harmless murmurs must be assigned to
the systolic aortic. Not only may this *bruit*
be heard when no disease of the aortic valves
exists; but even when this is not the case, it

still may be, and often must be, of very little import that a slight obstruction should, in an almost imperceptible manner, stand in the way of the transit of the blood from the heart into the aorta. So long as the obstruction is only sufficient to throw the blood into vibration, and does not require the exercise of any unusual force to overcome it, its presence is of no consequence. It is only serious in so far as experience teaches that it has a constant tendency to increase : and in proportion as the valve becomes more diseased, so does the chance of its permitting regurgitation also increase, and a regurgitant aortic murmur is a much more serious matter than a systolic one. The comparatively small importance of a murmur at the base of the heart heard along with the first sound is explained by the mechanism by which it is produced. The blood is propelled through the aperture with the whole force of the ventricle; and it requires a nice adjustment of the tenacity of the fluid passing from the hollow cavity into the smaller tube to produce the result that the motion should be effected noiselessly and evenly. It must be self-evident that mere rigidity of the valves preventing their falling back quite smoothly

against the arterial walls, or a very small frag-
ment attached to their edge, or a slight roughness of the surface of the artery itself, may each
in turn throw the blood into vibration. This
tendency will of course be much increased if the
tenacity of the blood be at all diminished by
disease, just as we know that when the thin-
ness and poorness of the blood have reached a
certain point, the murmur is produced without
any change in the solid structures at all.

There are other circumstances, however,
which give it an importance quite distinct
from its relation to enlargement of the heart.
In the first place, if the murmur can be properly
localised at the aortic valve, so as to give
pretty conclusive evidence that the disease is
in the valve itself, and the patient has had
one attack of rheumatic fever, to which therefore the damage may be fairly traced, we may
feel sure that another attack of rheumatism
would increase the mischief. Not only so, but
the reparative process which goes on tends for
some time after the attack to alter the form
and elasticity of the valve, so that the murmur
is very often quite altered in character, and a
slight obstruction becomes a more serious one,
or even a cause of regurgitation. In the

second place, if the disease be of a degenerative kind, we may be almost certain that it will go on increasing without any necessity for the supervention of more acute symptoms. In the third place, there can be no doubt that if degeneration can be traced at the commencement of the aorta, the same condition may be found anywhere else throughout the arterial system, and may there lay the foundation of ultimate serious disease. We cannot disregard the warning of head symptoms occurring in connexion with a systolic aortic murmur, however slight—not that the murmur necessarily indicates any such interference with the circulation as could of itself serve for their explanation, but that it suggests the possibility of arterial degeneration within the cranium with its possible consequences of apoplexy or softening of the brain.

Next in importance stands a systolic murmur at the apex indicating mitral insufficiency and regurgitation. In a former chapter I have mentioned my own experience of the possibility of life being very much prolonged when the affection is simply of this character. The blood during the ventricular systole presses on this aperture with considerable force,

and a very small amount of deposit on the
edge of the valves is sufficient to prevent
their complete adaptation: here a murmur is
most easily produced because the regurgitant stream, however small, is impelled
against the blood current entering the auricle
from the opposite side. The progress of
this form of disease is chiefly dependent on
two circumstances — viz., the actual quantity
of blood which is allowed to pass in the reverse
direction, and the demands made on the circulation by the amount of physical exertion habitually employed. Necessarily, too, the same
tendencies which have been mentioned as
leading to a progressive increase of the valvular imperfection at the commencement of the
aorta must operate also at the auriculo-ventri-
cular aperture.

It is unnecessary here to repeat the points
on which a correct diagnosis of mitral disease
must rest. The intensity of the murmur may
very often be taken as a proof of the extent
to which regurgitation is permitted, and of the
consequent danger of enlargement of the heart.
But we must remember that mere loudness of
sound is not always to be regarded as an index
of the degree of disease, as its intensity depends

on a combination of several causes, each of which may in any particular case exercise a greater influence than the rest. This is especially seen in cases of almost every day occurrence in which mitral disease, still of recent date, commenced during an attack of acute rheumatism from which the patient has not yet fully recovered. And as he has been reduced to a comparatively bloodless state by the febrile disorder, the murmur may suggest the idea of very extensive lesion, because the sound is intensified by the anæmic condition. A few weeks or months pass over, the health is re-established, and the intensity of the murmur is very much diminished — in great measure, as I believe, because the blood has returned to a condition of health. At the same time possibly the deposit is partially absorbed, and the imperfection of the valve considerably reduced at first; only to be again increased after a time.

Sooner or later, almost unavoidably, either as a result of the insufficiency having been considerable from the first, or of its having become so by increasing disease, the want of an adequate supply of blood is felt in the systemic circulation in consequence of its

M

return through the imperfect valve. This increased demand calls forth an increased propulsive power in the muscle of the heart, or in other words, hypertrophy begins. The valvular imperfection becomes more marked and more injurious as the enlargement of the heart proceeds; and this is explained by our occasionally finding mitral regurgitation quite distinct in a largely dilated and hypertrophied heart, when there is no appreciable disease of the valve. The imperfect adaptation is caused to all appearance in such cases either by mere changes of dimension in the cavities, or by some shortening of the *cordæ tendineæ*, or *musculi pectinati*. And if it be possible that a murmur should be produced in a valve apparently free from disease, by such a cause, the same circumstance must operate much more powerfully, we should imagine, where roughness or stiffness already produces imperfect adaptation of the valve surfaces to each other. Whether this be so or not, it will be acknowledged by every one who has watched the progress of such cases, that serious symptoms are only seen after changes in the size and capacity of the organ have occurred. It seems to me that the possibility of the existence of mitral murmur in the absence of true valvular lesion has not been sufficiently insisted on by systematic writers. A case has been already alluded to in an earlier part of this volume which is probably not an isolated one, of a mitral murmur persisting for some time, and then disappearing altogether, though there is no reason to believe that a valve once seriously damaged by inflammation ever recovers its perfect power of adaptation.

In chronic cases the fact has come too frequently and too palpably before me to leave any doubt in my own mind that many, very many times in old enlargement of the heart, when the pulse becomes feeble and intermitting, and the heart's action perfectly irregular, mitral regurgitation takes place, even though the murmur cannot be heard. In all questions referring to the mitral valve we must therefore remember that a time may come, when a previously distinct murmur ceases to be audible in the tumultuous action of the heart, though the pulse proves that regurgitation still goes on.

Very different in effect is the aortic regurgitant murmur from the systolic one at the same valve, as it is invariably attended from a comparatively early date by enlargement of the

heart. Its influence in this respect is much
greater than that of mitral disease alone.
From the earliest date of its existence the in-
completeness of the action implied in what is
called the systole of the heart—the immediate
return of the blood from the aorta into the
cavity from which it has just been expelled,
stimulates the muscle to act with increased
vigour; and it seems almost impossible that
such a state of things should exist without
enlargement resulting from it. It is consequently of no small importance in prognosis to
recognise its presence. It must be unnecessary
to refer to the rules for distinguishing a double
aortic murmur from a double friction-sound.
Any one who pretends to a knowledge of heart
disease, ought to be able readily to discriminate them. It is of more consequence to point
out what is not perhaps so generally acknowledged, that the point of greatest intensity,
which is of course taken as the index of the seat
of its production, is very often not at the base
of the heart, when the cause of the murmur is
disease of the aortic valve. It is by no means
uncommon to hear the sound much more distinctly at the apex than at the base, and when
in place of being double, regurgitation alone is

heard, a hasty inference may lead to the conclusion that the disease is mitral and not aortic.

The explanation of this phenomenon is not difficult, and the rules for correcting the inference are very simple. When disease has produced such a degree of roughness of the valve that the blood is thrown into vibration by passing over it in returning into the ventricle, the point of greatest intensity would naturally be just below the valve; but if the blood itself, in consequence of its watery state, exhibit any unusual tendency to functional murmur the loudness of the sound will necessarily increase to a certain distance from the point at which the vibrations commence. It must also be remembered that the returning current is not propelled with any great force, as it depends chiefly on the resiliency of the arteries; consequently when there is no great amount of roughness, the blood may pass quietly over the imperfect valve, and only be thrown into vibration when it encounters the onward stream flowing through the mitral valve, or when it impinges upon the unequal surface of the ventricular cavity. These are, I believe, the two chief reasons why aortic

murmur is sometimes heard loudest towards
the apex. But the sound so produced has not
the remarkably local character of mitral regurgitation; it is also diastolic and not systolic,
and the observer's finger placed at the site
of the apex-beat on the chest, will at once
correct any false impression. A systolic murmur
may begin a little before or a little after the
heart impinges on the chest; but it terminates half way between two beats. A diastolic
murmur, on the other hand, begins half way
between the two, and terminates immediately
before the following beat. It is just possible
that the diastolic sound heard loudest towards
the apex may be produced by the onward current
of the blood through the mitral valve. Practically such a sound is exceedingly rare, and it
is always accompanied by a loud mitral systolic
murmur. Indeed it could not well be otherwise: the onward flow of the blood through
that valve is slow, and the current weak, while
the returning stream when regurgitation is permitted is propelled with the whole force of
the ventricular systole. Nor does the systolic
murmur depend upon roughness of the valve
so much as upon the inversion of the current,
and consequently an imperfection which would

not produce vibration in the direct current of the blood even if its force were very much greater, may very readily produce a regurgitant murmur. On the other hand we cannot imagine a valve so rough as to give rise to a murmur when the blood is flowing through it during the diastole, and yet closing so perfectly as to admit of no regurgitation during the systole. These considerations have in my own experience served perfectly to distinguish between aortic regurgitation and any form of mitral disease.

Allusion has frequently been made in the preceding pages to alteration in the size and capacity of the heart, the true *morbus cordis* of older writers. Since the practice of auscultation has taught us to trace the earlier antecedents of this condition, we are too apt to lose sight of its importance. Nearly all the evils of valvular lesion and of adherent pericardium are traceable through changes in the muscular structure, and until such changes are established the patient seems scarcely to suffer in any marked degree, however loud the murmur may be. They consist either in increased thickness of muscle or increased dimensions of cavity, the latter being oc-

casionally accompanied by muscular atrophy.
The correctness or incorrectness of our prognosis depends in my opinion very much on a
just appreciation of these conditions when present; and there are certain fallacies which it is
well to bear in mind.

A person with a deep chest, and considerable power of lung-expansion may have a
heart of enormous size, while no direct evidence of its magnitude can be obtained. It
sinks down between the lungs, and becomes
so entirely encased by them, that its impulse
against the chest wall cannot be felt, and no
increase of dulness on percussion can be
traced; over the greater part of the præcordial
space the breath-sound may be distinctly
heard; yet the heart may be very greatly
increased in thickness, with cavities of very
large dimensions. In what used to be called
concentric hypertrophy especially, when the
walls alone are thickened without any marked
enlargement of the cavities, the probability of
the change being overlooked is very great. The
best corrective of this mistake is generally to be
found in the state of the pulse. A heart simply
hypertrophied without enlargement of its
cavities always acts with considerable energy,

and gives a sharpness or a thrill to the pulse, according to the state of the aortic valves. But when the cavities are also enlarged, a time very often comes in the history of the case when in spite of the thickness of the muscle it is unable to carry on the circulation efficiently in opposition to the increasing obstacles, and the heart's action becomes intermitting and irregular; the pulse then assumes a character of feebleness which is very apt to deceive.

Again, when the cavities are dilated, and the muscular wall thin or atrophied, it is very difficult to say whether the condition of the heart be not that of ordinary degeneration. The question perhaps seems very immaterial, if the power fail, what has been the cause of the failure? But in truth it is not so, at least in the earlier stages of the affection. The arrest of muscular degeneration must be attempted by means wholly different from those measures which would be recommended to obviate the consequences arising from dila-tation. And there are certain events more likely to happen in one condition than in the other. For example, sudden death is very often closely related to fatty degeneration, while passive congestions and œdema are the more constant results of dilatation.

In the uncertainty in which the observer is often left by a physical examination of the heart as to its actual condition, much may be learned from the past history of the patient and a consideration of the probable changes which are to be anticipated as the consequences of special diseased states. Subsequent experience has only confirmed the conclusions formed many years ago from a consideration of the *post-mortem* records of St. George's Hospital. A single systolic murmur at the base of the heart is generally followed after an interval, more or less protracted, by simple concentric hypertrophy. It indicates merely some obstruction to the exit of the blood from the ventricle, and it seldom happens that when there is no regurgitation, the blood in any degree stagnates in the cavity so as to bring on dilatation. This single systolic murmur is more frequently the consequence of atheromatous deposit than of inflammatory action, and atheroma of the aorta is constantly met with when this condition of the valve has existed for a time. Along with the spread of the deposit there is a loss of elasticity in the artery, which not only causes an increase of hypertrophy, but very generally also leads to an accumulation of blood in the heart, and is attended by a certain amount of dilatation. There are, of course, other causes which contribute to the development of simple hypertrophy, but they may be all summed up in the one expression of the existence of some obstruction to the onward flow of the blood. One only deserves special mention, because it is not very easy to understand how it leads to this result, viz., Bright's disease of the kidney. I have seen it most marked in cases in which there was no dropsy, and therefore no evidence of retardation of the circulation in the systemic capillaries. Various ideas have been suggested in explanation of the difficulty which it is evident the heart has to overcome, but none of them are to my mind perfectly satisfactory.

Aortic regurgitation is specially attended by the combination, in the highest degree, of hypertrophy and dilatation. After each systole the blood flows back from the aorta into the ventricular cavity, distending it with a double quantity of blood derived from two opposite sources, and pressed upon by the whole weight of the column of blood in the vessels.

This constantly acting tension gradually produces an increase in the dimensions of the cavity, while the necessities of the system demand that there shall be some counterpoise to the retardation of the blood-current consequent on its perpetual regurgitation. Such a counteracting force is only to be found in an increased thickness of the muscular tissue; and consequently in regurgitant disease we find the greatest development of hypertrophy with dilatation.

Mitral regurgitation may exist for many years in a moderate degree, without causing marked alteration in the size of the heart. So long as no unusual demand is made upon it the individual may be sufficiently nourished by a diminished supply. The circumstance that a small part of the current deviates from its proper course through the aorta, and a less quantity of blood than usual is propelled towards the systemic capillaries is not felt by the nervous system, and therefore no excitement of the heart's action takes place. Still there is a certain stagnation of the blood current, and the fluid accumulating behind the curtain of the mitral valve distends the ventricle rapidly after the systole is over, and very generally some dilatation may be traced. As soon as the effect of diminished supply comes to be felt, which happens sooner or later in each case, according to its particular character, then the action of the heart increases in energy so as to propel the blood more quickly and with more force, and thus to compensate the smallness of the quantity supplied; and as a necessary consequence of increased action hypertrophy ensues. Consequently, in cases of mitral regurgitation dilatation is the rule, and in the majority of cases this dilatation is accompanied by moderate hypertrophy. Exceptions naturally will be found in cases in which, either from extensive disease of the valve or other causes, over-action began early and continued throughout the history of the case, when simple hypertrophy may alone be found after death. In fact, whether with reference to mitral or to aortic disease, the extent of the muscular increase will depend chiefly on the existence of over-action during the systole, while the enlargement of the cavities is more directly traceable to the tension of the blood during the diastole.

Adhesion of the pericardium is another cause of change in the muscular structure of the

heart. Occasionally an attack of pericarditis
passes off, leaving the two surfaces pretty free,
but very frequently some amount of adhesion
takes place between them. Its extent is, of
course, very different in different cases. Sometimes only one or two bands pass across from
the one to the other, or the two surfaces
contract slight adhesions at the points where
the membrane is reflected from the wall to
the sac: sometimes the one is universally adherent to the other. Between these extremes
every possible gradation has been observed,
and it is found that they lead to very different
results. A very slight adhesion may produce
no appreciable effect; and the same result may
follow from an entire union of the two surfaces
together. Indeed it was assumed, at one time,
that this was the most desirable issue of an
attack of pericarditis, because the more complete conditions of recovery, marked only by
a white patch on the surface of the heart,
were then unknown. Complete adhesion, however, is found very frequently associated with
dilatation and atrophy. The muscle seems, to
a certain extent, to be deprived of its power, it
acts more feebly, and stagnation of the blood
takes place within the cavities, ending in their

dimensions being increased, while the muscular tissue is wasted. Partial adhesions also interfere with muscular movement, but rather excite increased action : and as the blood at the same time tends to accumulate in the cavities, a mixed condition of hypertrophy and dilatation generally results.

Such considerations derive all their importance from the difficulties occasionally attending the diagnosis of disease of the heart. It is very easy to point out sources of error, to which we are exposed in attempting to form correct conclusions, and it is not always easy to show how they are to be avoided. But, while on the one hand, there can be no doubt that the probability of error is least where the ear is most perfectly trained; yet, on the other, the evil consequences of disease are more constantly witnessed when the history of the case, and the general symptoms, most distinctly point to chronic organic changes, and the revelations of the stethoscope are of least importance. In the early stages of disease, it is of much value to be able to detect with certainty slight deviations from health, because at that period general symptoms are very apt to mislead. In the generation just passing away many a

young female suffered a martyrdom because
she was subject to palpitation, and the unedu-
cated ear of the medical man detected a *bruit*
which he was unable to refer to its proper
source in an anæmic condition of the blood.
From such a mistake proper training is alone
capable of guarding the observer. A careful
consideration of all the circumstances is quite
indispensible to correct diagnosis, and the most
accomplished stethoscopist may fall into error
if debarred from an enquiry into the history of
the case, and the sensations of the patient.
At the same time, it must not be forgotten
that while disease is yet in its early stage,
the discovery of its true nature, and the estimate of the evils which may arise from it,
must in chief measure rest on the results of a
physical examination.

When, on the contrary, the progress of the
case is marked by very considerable lesion, the
stethoscopic observations must always be cor-
rected and modified by reference to general
symptoms. Not only do we find, as already
pointed out, that a mitral murmur once very
distinct has disappeared, while yet the pulse
proves that regurgitation is going on; or that
a valve perfectly free from disease admits

of regurgitation, because its relations to the heart - walls are modified by alterations in thickness, and an accompanying change in the dimensions of the cavity ; but we find, too, that the position of parts is so changed that a perfectly false interpretation may be put upon a murmur which, when heard under ordinary circumstances, would from its locality alone be held to be conclusive as to the valve affected. A time generally arises in all diseases of the heart in which the diagnosis becomes more or less obscure, and the part of wisdom is to abstain from pronouncing a confident opinion. Then the labour of the circulation and the difficulty of breathing, the œdematous lungs and the anasarcous legs make it a matter of little moment whether the mitral or the aortic valves are chiefly at fault, and the main consideration is how far the heart's power may be sustained or regulated, so as to enable it to overcome the difficulty.

CHAPTER IX.

THE TREATMENT OF DISEASE OF THE HEART.

Principles of treatment — Stethoscopic indications — General symptoms — Retardation of the Blood-current — Its causes — Its effects — Albuminuria — Local congestions — Dropsy — Bronchitis — Congestion of Kidney — Of Brain — Deterioration of Blood — Rules of treatment.

THE treatment of disease of the heart is a subject of great importance and peculiar interest. Correct diagnosis is absolutely essential to success, and yet the result seems to be almost an accident. Remedies which give immense relief in one case utterly fail in another, when there can be no doubt about the character of the *bruit* heard or the form of disease present. The difficulties, however, are not greater in this than in other maladies, if due consideration be given to all the circumstances of the case. A broad and comprehensive view can alone deliver the practitioner from falling into the error of empiricism on the one hand or theory on the other. The blind reliance upon symptoms as the only guide to treatment, which for so many ages formed the sum of medical knowledge, and gave such overwhelming importance to mere experience, has passed away, and is succeeded by an honest and persevering effort to trace symptoms back in all cases to their true cause. That this effort should be invariably successful is more than could be looked for, but the measure of success which has attended it, has turned aside the current of observation too much from that class of symptoms which more naturally catch the eye and arrest the attention, and concentrated it on the unseen causes on which they depend, and the minute variations of structure and tissue which examination after death reveals.

As a necessary consequence, treatment has followed in the same direction, and it is not the cough or the shortness of breath, or the amount and character of the expectoration, which guide the practitioner in the selection of remedies, but the knowledge which he has obtained by a physical examination of the chest that there is disease of the heart, tubercle, or bronchitis present as the cause of these pheno-

mena. That they are important elements in
the calculation is undeniable, but I suspect
that many of the little helps to convalescence
which were familiar to our predecessors are lost
sight of by us in our neglect of the accessories
of important disease.

It is well known that when the heart is
affected the train of symptoms which follows,
is chiefly dependent on concomitant changes
in other organs, and these I have ventured to
assert are not traceable till some change in its
dimensions has taken place. It is, therefore,
not very material to the health of the patient
in the earlier stage of his disease whether the
murmur be heard at the base or apex of
the heart, and whether the interpretation put
on that murmur be exact or not. So long as
he is not harassed by unnecessary or injudicious treatment, no harm can come of any false
impression received concerning his case. But
as time advances deviations from the normal
dimensions of the heart commence, and such
changes are unquestionably very much influenced by the actual condition of the valves.
General conclusions drawn from other sources
must consequently be modified in conformity
with the result of stethoscopic examination,

and it is of no small moment that the information so derived should be trustworthy. We have already seen that the blood-current may pass in a backward direction when the heart is acting tumultuously and irregularly, while this movement is not revealed by the stethoscope; so that one who judges by general symptoms alone, may be more correct in his impression of the case and his idea of treatment than a man who relies solely on auscultation.

Correct diagnosis is clearly impossible if all the phenomena of disease be not taken into consideration; and in a much greater degree successful treatment depends, as it does almost exclusively, on a comparison of all the attendant symptoms with each other, and with that which we believe to be their exciting cause. That this should be so is a necessary consequence of the circumstance that if we exclude simple variations of intensity, the only remaining conditions which render one example of the same disease unlike another are shortly summed up under the term "symptoms." I may appeal to the consciousness of every practitioner whether any two cases are exactly alike, except when the whole powers

of life are subdued by some overwhelming
attack. I speak of such conditions as the
collapse of cholera, rupture of the stomach into
the peritoneal cavity, the prostration of malig-
nant scarlet fever, and so on. In all other
cases the minor adjuncts in their totality,
greatly modify an opinion based solely on a
consideration of the actual condition of disease
on which they are engrafted.

In their application to disease of the heart
such views of treatment teach us that we have
not only to investigate carefully the condition
of the organ itself, but take a review of all the
accidents which have been engrafted upon it.
Feebleness of the circulation for example is a
great and important fact on whatever causes it
may depend. However distinct the evidence of
hypertrophy of the heart, and consequent increase of power, the fact remains ; an insufficient
supply of blood is driven forward through the
arteries. To overlook this, and to rest upon
the merely physical signs which indicate very
clearly the existence of enlargement, is to miss
one of the most important guides in practice.
But it is manifest that as a weak circulation
may depend on a variety of causes, its remedy
will not always be the same. When the heart

is enlarged, and its walls thickened, there must
be regurgitation through the mitral valve when
the pulse is decidedly feeble. In most cases
we are left in no doubt about this inversion of
the current because the mitral murmur is distinctly audible; and when it cannot be traced
by the ear its existence must be assumed in
such a case. To endeavour to stimulate a heart
in which hypertrophy is more marked than
dilatation must only lead to further injury;
but in proportion as dilatation preponderates
and hypertrophy is less marked, so does the
power of the heart diminish until such a
degree of dilatation is arrived at, that an un-
naturally large organ gives the blood such
a feeble impulse as to represent the pulse of
mitral regurgitation. The same effect may
also be produced by a heart which is rather
below the normal dimensions, and has become
the subject of fatty degeneration. No one, I
think, could anticipate that the same treatment would be applicable in three conditions
so unlike each other, though all are marked
by the same feature,—one which is unquestionably a most important element in the causation of those secondary disorders so constantly
associated with disease of the heart.

The importance of this subject cannot be too strongly insisted upon. Secondary congestions are rarely owing solely to the increased impulse communicated to the blood-current by an enlarged heart. Without some retardation of the current we do not meet with that local stasis of the blood in the capillaries which especially marks the downward progress of cases of cardiac disease. My readers will perhaps picture to themselves cases too often occurring in practice in which this view seems to fail; cases in which simple hypertrophy of the heart with disease of the kidney, marked by a firm and rather hard pulse, exhibit all the symptoms which have just been referred to obstructed circulation. They do not seem to me to form any real exception to the rule. The very fact of enlargement having taken place proves to my mind that, whether we are able to give a full explanation of the phenomenon or not, there must have been some difficulty to overcome. The subject is an obscure one, and if we cannot dogmatise upon it, the theories which have been suggested are also not very conclusive. We do know indeed, regarding Bright's disease, that very many of the same local congestions occur during its progress which are usually found in connexion with impeded circulation, and it is quite reasonable to conclude that they are the true antecedents of muscular enlargement. But the explanation is not altogether satisfactory. Not only does the history in many such cases fail to give any indication of early pulmonary congestion, but a further difficulty presents itself in the fact that the hypertrophy is very constantly found in connexion with those forms of kidney disease which are not necessarily marked by symptoms of congestion at all. On the whole, I think, we must fall back on the hypothesis that the altered blood in kidney disease is the link which unites it with enlargement of the heart, when the two are found together. And, further, that the local congestions are themselves, less dependent on the state of the heart than on that of the kidney.

We cannot now enter on so large a subject as that of the ordinary associations of albuminuria. Suffice it to say that, in a certain number of cases, anasarca is a prominent symptom, and occurs early in the history of the disease, while in others it is late in its appearance, and does not form one of the

distinctive characters of the lesion, being in
great measure dependent on accidental complications. In both forms, however, the blood at
length becomes diseased and impoverished by
the failure of the functions of one of the great
emunctories, and the constant drain of its
albuminous principles; and then the exudation
of serum readily takes place. Alike into the
serous cavities, and upon the surface of the
mucous membranes, the effusion may be traced.
To such a result the existence of a certain
amount of congestion, or arrest of the blood
in the capillaries would seem almost essential;
but its character is passive, not active: the
internal organs are not gorged with blood,
but may rather be said to be œdematous; and
in but a few exceptional instances are the
phenomena called inflammatory ever found,
e. g., the slight coating of lymph on the surface
of the pleura or pericardium.

The presence of albuminuria is of immense
importance, whether by its permanence it indicate the actual existence of Bright's disease, or by its occasional occurrence we are
taught that the organ is only congested,
inasmuch as this state may lay the foundation for future degeneration. In several

particulars an analogy is established between
the sequelæ of cardiac and renal disease, but
the inherent difference between the primary
affections makes itself felt in all the details of
their consequences and terminations. Generally
anasarca is the common consequence of both,
and yet in the condition of the œdematous legs,
their tension and tenderness, and their appearance, no less than in the aspect of the sufferer,
are differences so marked that they cannot fail
to arrest the attention of the experienced observer. A liability to bronchitis and excessive
bronchial secretion is, not less than general
dropsy, a feature common to both diseases.
But the œdematous lungs of Bright's disease
differ as widely as possible from the gorged
and perhaps apoplectic lungs found in disease
of the heart. In head affections, too, there is
a similar general resemblance along with the
existence of equally marked individuality of
character. These facts sufficiently explain the
importance attaching to a distinct recognition
of each disease, when combined in the same
individual.

No two organs react so constantly on each
other in disease as the heart and kidneys. To
such an extent does this occur that it is really

seldom possible to class cases of dropsy as distinctly cardiac or distinctly renal—terms which are in constant use with those who are not in the habit of studying the relations of diseased organs to each other. In a few cases of granular degeneration the dropsy appears without any trace of heart affection, and in a still smaller number it commences as a sequel of disease of the heart before any congestion of the kidney has been discovered. But the number of such instances is very limited in comparison with those in which disease of both organs combines to produce the effect. Similarly, in cases of hemorrhagic apoplexy, the combination of hypertrophied heart and granular kidneys is much more commonly found than hypertrophy alone, though it is not very easy to explain the influence of diseased kidney on this form of brain disorder.

In further considering the consequences of heart disease we cannot fail to be struck with the circumstance that they are by no means constant, and are not always the same in different individuals. Many secondary causes come into operation which determine their special character, and not unfrequently we

are quite at fault in attempting to trace their
causation. In one who is exposed to changes
of temperature, and whose mucous membrane is
liable to be excited by atmospheric vicissitudes,
the change from chronic bronchitis to passive
congestion with abundant secretion, in consequence of mitral regurgitation, is most natural;
and though perhaps less frequently, yet not
less certainly, may the same event happen
when the blood accumulates in the left cavities from aortic insufficiency and regurgitation. It is not so clear why in another person
the strain falls more heavily on the kidneys,
or why in others the liver or the brain suffer most from congestion when the cause is
to be traced to the general circulation. And
while in the brain the effect may be partly
accounted for by degeneration of arteries, it
is not unnatural to assume that though not
so easily proved by the rough modes of investigation we are in the habit of employing, yet
there must be in the organ attacked some internal vice, natural or acquired, which directs the
selection, when one suffers more than another
in consequence of obstructed circulation.

To none of the complications of heart disease am I inclined to assign a more important

place than to the presence of serous effusion. It is never seen in the early stages, whether the change of structure be dependent on inflammation or on degeneration, and can only be traced after enlargement has already occurred. At the same time it does not necessarily follow when enlargement is of the most formidable description, and is rarely found to any extent without the intervention of some other circumstance. As has been already remarked, the only idea we can form of the cause of muscular thickening is the existence of some obstacle to the onward movement of the blood; and when the retardation is of such a character that the blood accumulates in, and distends the cavities, dilatation is combined with it. The enlargement then is only a counterbalancing force, and so long as it is no more than just sufficient to overcome the obstacle, there is no decidedly greater reason why dropsy should occur after its existence than before it. But the balance is not always equally adjusted, and the increase in the dimensions of the cavities aggravates the valvular imperfection, and adds to the delay in the onward movement of the blood. Consequently cases are occasionally met with in practice in which the damage to the heart, and that alone, seems to cause effusion of serum. Unusually great or long-continued exertion produces œdema of the ankles: perhaps rest in the horizontal posture at once removes all evidence of its existence, but the patient is no longer safe from its recurrence; and it does return on the slightest provocation.

More commonly some other event combines with the cardiac imperfection to produce that local stasis of the blood, by which its serum is made to exude through the walls of the minute vessels. Two such circumstances deserve especial notice; first, the occurrence of bronchitis; and, secondly, the existence of congestion of the kidney apart from fully developed disease of that organ. When, as usually happens, the retardation takes place in the left ventricle, in consequence either of mitral insufficiency which causes the deviation of some portion of the blood from its onward course, or of aortic regurgitation which permits more or less blood to flow backwards out of the artery once filled by its stream, there must be almost of necessity an excess of blood in the lungs. It is only when the supply is very scanty, and the amount poured into the right

side of the heart by the *rena cava* is insufficient adequately to fill the pulmonary arteries,
that the tension is not felt. But a person
so circumstanced must be exceedingly liable
to bronchitis on the least exposure. The inflammation of the mucous membrane again,
renders it more difficult for the large quantity
of blood in the lungs to be properly exposed
to the influence of oxygen, and a fresh cause of
retarded circulation is immediately introduced,
giving rise to more congestion, and increasing
the bronchial inflammation; the two conditions thus react on each other, and constantly
augment the difficulty of the circulation.
Indeed the congestion of the whole lung tissue
from the excess of blood which it contains must
be almost of itself sufficient to excite symptoms
of bronchitis, without exposure to cold or to
any other of the ordinary exciting causes.

An attack of bronchitis originating in
this manner is much more obstinate and
lingering than it would have been if the
patient had been in health. Its duration,
and the slow changes in structure caused
by its occasional recurrence, lead ultimately
to a condition in which pulmonary congestion is constantly present, and · the blood

cannot pass through the lungs at its ordinary rate. This new cause of retardation is propagated backwards through the right cavities to the venous system, and engorgement of veins with imperfect propelling power in the heart brings the flow through the capillaries almost to a stand-still: exudation of serum takes place, and the absorbing power, limited as it generally is by the greater or less rapidity of the circulation, is reduced to a minimum; œdema begins, which can only be relieved by quickening the pulmonary circulation. To this object remedies must be mainly directed, and so long as imperfect oxydation of the blood continues, and the whole surface of the body presents a purplish colour, with marked blueness of the face and extremities, little hope can be entertained of any decided diminution of the dropsy. In vain we seek to get rid of fluid by free discharges from the intestinal canal, or the administration of diuretic remedies. The water eliminated through these channels is soon made up by the process of absorption, and while the impediment lasts, serous exudation into the areolar tissue must go on.

Congestion of the kidney, again, is a very

frequent effect of disease of the heart, and almost before it can be discovered acts as a new cause of dropsical exudation. For a certain period, persons with heart disease present no trace of albumen in their urine, provided the kidneys are free from disease. Ultimately the balance of the circulation is very seriously disturbed, and congestions of all internal viscera are prone to be developed. To me it seems not improbable that they are secondary in most cases to the pulmonary obstruction just alluded to, and its attendant venous congestion. When the kidney is the organ affected the urine becomes scanty and loaded, and a trace of albumen is discovered : if it be possible either to moderate the general obstruction to the circulation, or by means acting more directly on the kidney to soothe irritation, and relieve congestion there, the urine may be once more secreted in its normal quantity, its turbidity disappearing as well as the trace of albumen. But if the congestion go on, and still more if it be renewed on several occasions, slow inflammatory action is set up, the secreting structure is disorganised, and permanent albuminuria is established. We have now no longer to deal with a damaged heart causing disturbance to the circulation in distant organs, but with the much more formidable condition of combined heart and kidney disease, than which very few states are less amenable to treatment. The huge anasarcous legs become daily more tense and loaded with serous fluid, the abdomen is distended with a similar effusion, and gradually the pleura and pericardium participate in the same condition ; the functions of all these organs being so seriously interfered with that life itself cannot go on.

The head symptoms which attend cardiac disease are also of great moment in the treatment of such cases. Here the condition is perhaps rather one of active than passive congestion, though probably both are combined in the development of the phenomena which in their totality constitute an apoplectic seizure. It not unfrequently happens that one who has long been the subject of disease of the kidney, and whose heart has become simply hypertrophied in connexion with albuminuria, at last suffers also from degeneration of the coats of the arteries : the vessels lose their elasticity, they can no longer bear the increased strain of a blood-current driven onward with considerable force by an hypertrophied heart, and they give way: hemorrhage immediately takes place, and the patient is suddenly deprived of consciousness, perhaps of life, as some persons never rally from the first attack. Such cases are of constant occurrence, both in hospital and in private practice, but they stand almost alone as representatives of congestions caused directly by hypertrophy. More commonly, I think, we may trace valvular lesion as one of the elements in the causation of such attacks. The impediment to the blood-current from degeneration of the arteries is not alone sufficient to produce that tension which ends in rupture; some other cause of retardation is needed to complete the congestion of the cerebral vessels. The impulse of the blood driven onwards by a heart acting with an abnormal power could not alone produce extravasation so long as the blood flows evenly and uninterruptedly through the brain. The obstacle to the return of the blood, may, as we have seen, be produced by imperfection of the valvular apparatus; the same degeneration which destroys the elasticity of the arteries is constantly found affecting the valves, so that by a natural and almost necessary consequence the two conditions are associated in the production of head symptoms. I have referred only to the complete apoplectic attack, but in minor degree local congestions may be present, and by their indications give warning of coming danger, when no seizure actually takes place. These form part of the train of symptoms which must be met by appropriate treatment in the management of all cases of disease of the heart.

Another consideration ought to be present to the mind of the physician in reviewing the various effects of disease, which seems to have a very direct bearing on the remedies he may select, and the general rules for guidance he may lay down in each case which presents itself to his notice. The blood itself cannot but suffer deterioration from the imperfect action of almost every internal organ, each of which is in turn liable to be affected by local conges-tion in disease of the heart. In so far as they act as purifiers of the blood, the partial suspen-sion of their function leaves a residuum which gradually accumulates in the system. The carbon is not properly removed by the lungs, the urea and salts are not thoroughly evacuated by the kidneys. Imperfect nutrition results from congestion of the liver and the whole

of the digestive apparatus. The effect is necessarily proportioned to the extent to which the
organs have suffered. It scarcely reaches the
degree of intensity seen in chronic disease of
the liver or kidney, when extreme emaciation
or uræmic poisoning serve so constantly to
indicate the limit which disease has attained.
We have already seen reason to believe that
pulmonary congestion is very generally the
first in the sequence, and I may add that it is
also the most marked throughout, and demands
more than any other the attention of the
medical adviser. The blood is dark, and
imperfectly freed from its carbonic acid,
and so long as this is the case it cannot afford
a healthy stimulus to the functions of life.

I will not venture to lay down specific rules
of treatment. My object in going into the
details which have just occupied our attention,
has been to indicate the direction which must
be followed with the view of ameliorating the
condition of the patient. A cure of the primary
lesion is so entirely beyond our power that the
principles laid down at the commencement of
this chapter seem to me to offer the only guide
which can be of any avail. The sufferer is
labouring under disease of that organ whose

function is to propel the blood through the whole frame. Careful inquiry must therefore be instituted to ascertain the exact nature of the disease, as it affects the heart's walls or the valvular apparatus, and the special treatment pursued must be modified in conformity with indications derived from this source. Still we must keep even more prominently before the mind's eye the secondary affections which have been produced, because on their relief so much depends. Let these disturbances but be removed, and the individual is once more restored to a condition of comparative health,—much below par indeed, and continually tending to deteriorate,—but, after all, only the state to which he has been gradually accustomed, probably, for years. That this is not always possible is not the opprobrium of medicine, but one of the necessities of an incurable disease; and while skilful treatment may greatly relieve, blind empiricism will probably aggravate all his sufferings.

To one or two conclusions from the view which has been taken of disease of the heart, I would crave the attention of my readers. A feeble pulse must be treated in a manner wholly different from one with a good deal of force and firmness, however analogous the cases may be in other respects. But we must not stop here. Till it is known on what the loss of power depends we are in no condition to prescribe for the patient. The heart may be hypertrophied and acting with much force, but driving the greater part of the blood backwards through the mitral orifice; or it may have greatly attenuated walls and large cavities; or its muscular structure may be degenerated into fat. Stimulation which may at least produce temporary benefit in the latter conditions, would only aggravate all the evils of the former; whereas digitalis by controlling over-action may give strength and regularity to the pulse. In attempting to relieve dropsical effusion, again, we must endeavour to trace the point in the circuit of the blood where the retardation of its current takes place, no less than the cause of obstruction. No effort to pump off the fluid will succeed unless this obstacle be first overcome. And here the condition of the kidney comes prominently forward. When albumen is present, if it point to a state of congestion only, prudence urges us to refrain from stimulating the organ to further excitement, and rather suggests measures calculated to soothe

irritation. If, on the contrary, actual degeneration exist, it may sometimes be necessary
to run the risk of producing further mischief
in giving stimulant diuretics; inasmuch as
temporising measures cannot restore the organ
to health, and relief can only be obtained by
increasing the amount of the kidney secretion.
I think it is consistent with experience to say
that the free administration of gin is not
unfrequently attended with the best results,
however much theoretically such a practice
might be condemned. In regard to the
pulmonary affection it is of importance to
ascertain whether the changes are only of recent
date, and imply mere congestion of the mucous
membrane, or whether permanent thickening
and opacity are present. Such changes are
usually best recognised by the appearance of
the sputa. It is of still greater importance,
however, to ascertain whether the lung is
choked with serous exudation, or is gorged with
blood. The treatment must differ in each case,
and it is not always very easy to distinguish
between them. But I need not go into further
details. One who has learnt what medicine
can really accomplish, and where its power
ends, and who takes the trouble to master the

details of symptoms, will not hastily prescribe unnecessary remedies, and will try to accommodate his means to the end which he seeks to accomplish. The presence of heart disease, while claiming some consideration on its own account, does not very much modify the plan of treatment applicable to any of those secondary disorders which occur during its existence.

CHAPTER X.

Peculiarity of Gouty affections — Changes degenerative — Spasm
and irregular action — Gout in the Stomach — Intermittent
pulse — Its causes — Treatment of Gouty affections — Dr.
Gairdner's views — Injudicious use of Colchicum — Conclusion.

In cases of gout, the series of symptoms
referrible to the heart, differs very materially
from those occurring in connexion with rheumatism, and exhibits certain modifications of
the ordinary progress of cardiac disease. As
has been already stated in the earlier part of
this volume, gouty persons are not generally
liable to inflammation of the heart. Sometimes
a deposit of lithate of soda may be seen on one
or other of the valves, but there is no evidence
that it has resulted from inflammatory action :
like the gouty incrustations on the ear it is only
one form of mal-nutrition attendant on the
diathesis. The valves themselves are commonly

atheromatous if any disease be present, and there is never any trace of the previous existence of pericarditis at all analogous to that which so often remains after acute rheumatism. The kidneys are very liable to become granular in the later stages of the disease, and possibly in association with the presence of albuminuria, recent lymph may be found either on the valves or pericardium, but not as the result of gouty inflammation. The changes which take place are of a degenerative kind. The valves and the muscular tissue alike give evidence of this fact, as atheroma is found on the one, or fatty degeneration occurs in the other. Occasionally, however, the kidney disease leads to the development of a certain amount of hypertrophy, whether valvular lesion be present or not, and then degeneration of the walls of the heart is not perhaps so likely to occur.

Gouty affections of the heart, as commonly understood, are quite distinct from organic change. The same unhealthy blood which in its transit affects one joint after another, now leaving the part first attacked, and then again perhaps returning to it after visiting some other limb, or some internal organ, is no doubt the medium through which the heart becomes implicated. The affection has, too, the same transitory character; sometimes characterised by spasm which lasts but for a very short time, and seems of such intensity that life must cease if immediate relief be not obtained; sometimes seen in irregular action, which may last for days; but in either case leaving no trace of its past existence in organic change. The presence of actual disease, here necessarily plays a very important part. A damaged organ is more likely to be attacked than a healthy one, and the symptoms are usually more severe and of longer continuance. Indeed, it appears that in the advanced stages of disease the irregular impulse given to the heart by an attack of misplaced gout, may remain unrelieved till the patient's death, just as happens where any other strain overpowers the rhythmical action of a diseased heart. It is especially in feeble or fatty hearts that this event is to be apprehended: when the muscular structure is unchanged, even though valvular disease be present, there is more hope of a return to regular action than when the walls are degenerated or the cavities dilated. It is not always easy to determine whether the spasmodic attack is seated in the heart itself or in the stomach. The patient is so prostrate by the agony which he suffers that he can scarcely give an intelligible account of his own sensations; and when the heart is in any way diseased, the presence of flatulent distension which usually attends gout in the stomach, as it interferes with its regular action, leads to the idea that it has been primarily affected, when in truth the symptoms there, are only secondary. The spasm seems to me to be very generally different from that of angina, the only other example of heart spasm with which we are acquainted; and while in the one sudden death is not unfrequent, in the other it is comparatively unknown. I think it is of importance to trace out if possible the exact seat of the pain, because the treatment will be somewhat modified by a correct appreciation of its real character.

Irregular action when due to stomach affection is less directly caused by it than when the symptoms are due to spasm. It is not always possible to trace the cause of irregular action of the heart. By no means confined to gouty persons, neither of necessity indicating any actual disease, it seems to be made up in most

cases of a nervous element and a blood element. Very many persons who enjoy perfect
health have an intermittent pulse. This defect
indicates an imperfection in the nervous power
which controls the heart's movements ; the organ
does not fully respond as it ought to do to the
stimulus. The intermission generally takes
place at considerable intervals, and recurs with
a certain amount of regularity. If from any
cause the intervals become shorter and more
variable in their duration, intermittent action
is very liable to merge into irregularity. Such
a condition of pulse seems to depend on two
causes—either the nervous system is lowered
by exhaustion, or some form of mal-assimilation renders the blood less healthy than before.
The nervous filaments may therefore respond
less readily to the stimulus, either because
their receptivity is impaired, or because the
blood is unfit for its proper uses in the economy.
We can very readily understand how the unhealthy blood of gout may assume exactly that
condition which unfits it for the proper stimulation of the heart's action, and so long as that
unhealthy character is retained, a heart free
from disease will beat irregularly. Why it
commences somewhat suddenly during the

gouty paroxysm we are not able to explain, any more than we can assign a reason why the gout comes out suddenly in the foot or elsewhere.

The subject of irregular action has not received the attention which is due to it, whether we have regard to the sensations of the patient or the prognosis to be formed. To most practitioners an irregular pulse conveys the idea of serious disease of the heart, and if by the stethoscope no valvular imperfection is revealed, the conclusion generally is that some degeneration of the walls or dilatation of the cavities must exist to account for its presence. That such is not necessarily the case I am fully convinced, because a pulse which to-day seems to be perfectly irregular in its beat, may to-morrow be quiet and uniform. Careful observation would probably prove that such irregularity was only a very exaggerated form of intermission, but I confess that it is not always possible to form a clear opinion on the subject until a return to normal action indicates that the disturbance was only functional. To many of my readers this observation may not be new, but I think the majority of medical men are not aware of the possibility of such alternations of regular and irregular action occurring without organic change.

In the treatment of gouty affections of the heart there are several considerations which ought to be always present to the mind of the practitioner. The urgent symptoms must be relieved by such means as will act most promptly and energetically, and the unhealthy condition of the blood must be improved, regard being had to organic changes which may have taken place in the heart or the kidneys. The heart affection is generally preceded by the retrocession of the gout from some joint where it had been previously located : what is called a metastasis has taken place, and whatever other measures be adopted, it is urgently necessary to induce its reappearance in some external part. Sinapisms and hot fomentations or other irritants must be freely applied to the feet, where it more readily displays itself than elsewhere. The effect of stimulants and carminatives, which very often produce a large discharge of flatus lend probability to the view that the heart is often only secondarily involved ; but frequently the attack does not pass off till there has been full time for their absorption into the system, and for their influence to make itself felt in the circulation. It does not come within the scope of this volume to do more than point out in a very general way the indications for treatment. To those who desire to learn the lessons of an extended experience I can heartily recommend Dr. Gairdner's very valuable treatise on gout, and I should scarcely feel myself justified in laying down rules which must so entirely coincide with his, except in his own words. His views regarding the administration of colchicum are such as will commend themselves to the judgment of every thinking mind, and must, I believe, find a response in the conviction of every one who has had much practice in its employment.

In no case will its injurious operation be more manifest than when an attempt is made by its means to relieve patients in whom the gout has attacked an internal organ. There is something so striking in its power of arresting the early attacks that both patient and physician are alike tempted to forget general symptoms, and have recourse to its repeated employment whenever gout lurking in the system finds for itself a point of attack. But when the whole blood is contaminated by a long continuance of habits of indulgence in spite of previous warnings, when the liver is congested and unable to secrete a healthy bile, when the kidneys pour out a pale and acid secretion in which the uric acid salts are eminently deficient, and the powers of the stomach and intestinal canal are so weakened that healthy assimilation is impossible, while the bowels are probably loaded with accumulated fæces, what ultimate benefit can accrue from checking the local manifestation? Far wiser is it unquestionably to aid nature in the elimination of the peccant matter from the system, a result invariably attained by the turmoil produced when a sharp attack of gout in the extremities occurs. The first evidence indeed that the paroxysm is coming to an end, is always the reappearance of lithic acid salts in the urine. Our endeavours must therefore be directed to restoring healthy secretion in such cases before we attempt to arrest the disorder. The acute stage must sooner or later pass off if wholly left to itself, and complete freedom can only be obtained by a return of all the functions to the condition of health.

Nothing perhaps stands so much in the way of recovery as disease of the kidney. Heart disease also exercises considerable influence because it so constantly causes local congestion and interferes with healthy function: but when those organs which in health preserve the balance between the formation of uric acid in the system and its excretion in the urine are themselves diseased, and their secernent action interfered with, and when the tendency to acid formation is abnormally excessive, it is extremely difficult to preserve the patient from an almost unbroken series of chronic and ill-developed attacks of gout. These points afford subject for deep thought to the careful practitioner. He will not satisfy himself with procuring temporary relief from suffering, because he knows that a tedious and trying illness will surely follow, and then he may find all his remedies fail of procuring the most transitory relief, and he may have to stand by, as a helpless spectator of suffering which it is too late for him to attempt to alleviate.

In conclusion, I may be permitted to say

that the short sketch which has been attempted
in the preceding pages has been the result of
much time and thought devoted to the subject
of disease of the heart. It would have been
very easy to give numerous examples in proof
of the statements which have been made, but
I have thought it better not to weary the
patience of my readers with details of cases.
Salient examples may be occasionally selected
for the purpose of proving particular points
in practice, but the convictions which force
themselves on the mind when devoted to the
study of any particular subject are produced
by the combined influence of many instances,
and cannot be verified by solitary examples.
It is very seldom that an *experimentum crucis*
can be given in medicine : the laws of causation are far too complex to be resolved by
individual cases, however simple. It is only
when an experience is very frequently repeated
that it assumes any real character of probability, and the relations of sequence and
causation must be judged of by the test of
varied circumstances and relations. If I have
appeared too dogmatic in assertion it is only
because a long experience has seemed to me to

offer a full confirmation of ideas, which first shaped themselves into form when I was, many years ago, studying the relations of heart disease in St. George's Hospital, and I shall be glad to find that they meet with confirmation from the experience of others.

THE END.

LONDON: PRINTED BY WILLIAM CLOWES AND SONS, STAMFORD STREET,
AND CHARING CROSS.